LTE Signaling
with Diameter

About the Author

Travis Russell has more than 35 years of experience in the telecommunications industry and is a recognized expert in Diameter, SS7, IMS, SIP, and communications security. He is the author of *Signaling System #7, The IP Multimedia Subsystem (IMS): Session Control and Other Network Operations*, Session Initiation Protocol (SIP), *Telecommunications Protocols*, and *The Telecommunications Pocket Reference.*

LTE Signaling with Diameter

Travis Russell

Mc
Graw
Hill
Education

New York   Chicago   San Francisco
Athens   London   Madrid
Mexico City   Milan   New Delhi
Singapore   Sydney   Toronto

Cataloging-in-Publication Data is on file with the Library of Congress.

McGraw-Hill Education books are available at special quantity discounts to use as premiums and sales promotions, or for use in corporate training programs. To contact a representative please visit the Contact Us page at www.mhprofessional.com.

LTE Signaling with Diameter

1 2 3 4 5 6 7 8 9   QVS   21 20 19 18 17 16

ISBN 978-1-25-958427-5
MHID 1-25-958427-5

The pages within this book were printed on acid-free paper.

Sponsoring Editor
Michael McCabe

Editorial Supervisor
Donna M. Martone

Acquisitions Coordinator
Lauren Rogers

Project Manager
Apoorva Goel,
Cenveo® Publisher Services

Copy Editor
Surendra Shivam,
Cenveo Publisher Services

Proofreader
Anshu Sinha

Indexer
Cenveo Publisher Services

Production Supervisor
Lynn M. Messina

Composition
Cenveo Publisher Services

Art Director, Cover
Jeff Weeks

There have been so many positive influences in my life, both personally and professionally. Only one has been consistent and loyal most of my life. This book is dedicated to that one person who has stood beside me through the good times and the bad times. To the love of my life, my wife Deby.

Contents

Acknowledgments

Oracle USA has been a great experience for me. I have worked for large companies before, but for the last 24 some years I have been working for a small company (Tekelec). When we were acquired by Oracle a few years ago, it was a breath of fresh air for me. I was given the opportunity to do what I do best, and it opened up new experiences. So many thanks to Oracle and all those I engage with on a daily basis.

Thanks to Doug Suriano and Susan McNeice at Oracle for reviewing the outline for this book and giving McGraw-Hill the green light. When Oracle asked for some references to review the outline, they were the first that came to mind. Doug for his technical expertise and vision, and Susan for her communications skills.

Thanks to Chris King for encouragement and providing me the opportunity to excel in my role as technologist and cyber security "guru." I use this term lightly as one never learns everything, and I know I have a lot to learn.

My wife deserves the Medal of Honor. She is not a morning person, so she is not as eager as I to get up and start the day while the sun is still rising. But through every project she has been by my side, encouraging me, pushing me through the writer's block, and just being there when I needed her most. Thank you Deby for your patience, understanding, and your love.

CHAPTER 1

Introduction

What This Book Is About

Diameter is a complex protocol. I have studied Signaling System #7 (SS7), Session Initiation Protocol (SIP), and countless other protocols throughout my career, and Diameter has been the most complex. There has been a lot written about Diameter networks, focusing mostly on the various network functions. There have also been some books that combine the discussion between some aspects of Diameter and more about the network.

That is why I wrote this book. I wanted to write a book about the protocol itself, how it works, and how it is used in various sections of the network.

This book is not about 4G networks. Nor is it about charging networks. This book is about the Diameter protocol, and its usage in all networks. The intention is not to focus on the network anymore than necessary for explaining the role of the protocol.

When I set about to write this book, I struggled with the organization. Diameter as a protocol continues to grow, and the documents covering this technology have become expanse. When writing the book on SS7, I remember sitting at the kitchen table with three big binders that constituted the entire protocol. Not the case with Diameter.

The 3GPP documents are numerous, and navigating through all the documents can be a daunting task. This is why I set about writing this book. To provide a single point of reference where one could get a basic understanding of the commands and their parameters, and the basics of the many interfaces.

I knew I could not capture every interface, and every command. My intent was to write about the principal applications, technicians, and engineers would encounter on a daily basis in the course of their work. I think I have accomplished that, but I know there will be more to add in future revisions.

You will see in the bibliography that there were many sources for this book. It would be impossible for me to document each application of Diameter in detail, so it is necessary to use these same references in your own studies of the protocol to get the details on how each application functions.

They are free, and accessible through a search engine on the Internet. For IETF documents, search by the RFC number. For 3GPP documents, search on "3GPP" and the document number (i.e., "3GPP" 29.372). You will find the 3GPP site listing all revisions of the document as your first choice usually. This I have found to be the best place to retrieve 3GPP documents because you get the latest revision, but keep in mind it may not be a fully ratified version. Those are usually found in .pdf format on the same site.

What Is Signaling

Deep in the heart of telecommunications networks, there is a technology that controls everything that happens within the network. Think back to the days of cord boards. An operator sat at the end of a connection, and controlled every interconnection. If you wanted to connect to another city, another country, you were connected through a series of cables and wires through these cord boards. The operator was the brains of the network, receiving information about the call from you the caller.

As time has evolved, the technology used in this control network has advanced. We started using tones that were carried over the voice circuit, and were used to communicate with other switches about routing for the call. These tones were very limited, as there is only so much information that can be sent using this method. Another problem with the tones is that they were easy to duplicate, and hackers quickly learned how to recreate these tones through "black boxes" and even whistles (anyone remember Captain Crunch?).

With the digital age came new technologies, and the signaling network was born. There were iterations we will not go into, but SS7 replaced the tones in the network, and enabled us to support so many new capabilities.

Wireless networks were made possible through the use of the SS7 protocol, because of its ability to share information throughout the network. But until this point, it was all about voice communications. Data sessions were still relegated to fixed circuits, and broadband connections. This changed when the 3GPP began defining the migration to an all-IP network for wireless networks.

Leaping ahead a decade or two, networks today have successfully migrated to an IP infrastructure, enabling a lot more control data to be shared between network elements. SS7 is no longer capable of supporting the new capabilities that wireless networks demand, and fixed line broadband and cable networks are in need of a more powerful control technology.

The SIP took over from SS7 ISDN User Part (ISUP) and became the standard for signaling for voice and multimedia sessions in networks that are all IP. SS7 remains in some segments of the network, and probably will for several more decades. But there was a need for a new signaling protocol to replace the aging Remote Access Dial-In User Service (RADIUS) protocol. That replacement is Diameter.

So signaling is what controls everything we do with our devices. Today, turning on your phone creates numerous signaling messages that travel through the network to verify your identity, exchange cryptology keys between network nodes, and even to identify where you are connecting and what you are sending. Signaling is the brains of the network. There are three signaling protocols that are used throughout the world today in telecommunications networks, each one supporting a specific type of session. SS7 is still widely in use throughout the world supporting 2G and 3G networks, SIP is used for voice communications in IP networks, and Diameter is the signaling technology for the next several decades in all-IP networks.

Experiences While Writing the Book

I am often asked when I have time to write books. I have a full time job that is sometimes very demanding, and I travel. A lot. In fact I am balancing my computer on my lap as I write this paragraph on a flight to Rome. So I thought it might be interesting to log when and where I do my writing this time.

Planes are horrible. The seats are crammed and there is no room to do much other than sleep and maybe read. No, I don't get the luxury of flying first class. I am usually at the back of the plane in economy. I have grown quite accustomed to it actually, but the ability to use my PC has limited my writing time while on these flights.

I was fortunate enough on a few flights to have a roomy seating arrangement where I was able to actually use my laptop and do some writing (like I am doing right now). These are the flights where I was able to write:

Etihad Flight EY102, JFK to Abu Dhabi

Etihad Flight EY470, Abu Dhabi to Singapore

British Airways Flight BA1550, Heathrow to Raleigh

Toronto International Airport

Dallas International Airport

American Flight 1414, Mexico City to DFW

American Flight AA110, Rome Italy

American Flight AA1349, Raleigh to Dallas

American Flight AA174, Raleigh NC to London Heathrow

American Flight AA974, Rio de Janeiro to JFK

American Flight AA916, Bogota Colombia to Miami FL

Qantas Flight QA93, Melbourne Australia to Los Angeles CA

Hotels are the best. I am not a big TV fan, so reading and writing is my favorite diversion. Especially in foreign countries where the TV programs are in another language. These hotels are where I did some of my best writing:

The Mirror Hotel, Barcelona, Spain

St. Regis Hotel, Abu Dhabi

Shangri-La Rasa Sentosa Resort & Spa, Singapore

Barbizon Palace Hotel, Amsterdam, Netherlands

Hyatt Regency Toronto, Canada

Marriott Grand Hotel Flora, Rome Italy

Doubletree Hilton, San Jose California

Marriott Hotel, Mexico City

Grosvenor, Victoria London, UK

Comfort Inn, Jonesboro, NC

Windsor Atlantica, Copacabana, Rio de Janeiro, Brazil

Vinoy Hotel & Resort, St Petersburg, FL

Hilton Cartagena, Colombia

The Langham Hotel, Melbourne, Australia

The Mandarin Oriental, Suntec City, Singapore

And of course, a lot of writing got done at home. My best times for writing are early in the morning. I am known to get up at 5:00 am and start writing until its time to go to work. Since I work from home, the commute is not very far!

CHAPTER 2

Introduction to Wireless Networks

Wireless networks have evolved dramatically over the last 10 years, offering a plethora of services today that we could not possibly have imagined just 10 years ago. These services would not have been conceivable had it not been for the technologies developed over the years to support the delivery of those services.

The standards bodies have defined an evolution for wireless networks that has taken us from 2G networks to 4G today (and soon 5G). Every generation of technology offers higher performance at the air interface, supporting faster and faster data rates between the device and the network. At the same time, these standards also define changes in the network architecture to facilitate faster speeds and a more IT-centric infrastructure.

The most important part of this evolution has been the move to an all-IP infrastructure; a departure from the tried and true time division multiplex (TDM) facilities used in earlier networks. This move to all-IP not only improves the performance of the network, it supports higher data rates in the network core while reducing costs.

This chapter outlines the evolution to an all-IP network and the technologies used along the way, leading up to the introduction of Diameter. This is not an exhaustive description of network evolution, just an overview. There are many books available that provide more details about the network architectures and the evolution to IP. Our focus will be an area not covered; the Diameter protocol itself.

Evolution of the Mobile Network—From 0G to 5G

The first wireless networks were found in limited supply. They were expensive, and the service was horrible. Only the rich and famous could afford a radiotelephone in their vehicle. A radio tower could support up to 300 simultaneous conversations, and the coverage area was quite large. This was before the advent of cellular technology.

Direct dialing was not even possible in these early networks; you connected through an operator who made the connection to a landline for you, then called you back when the connection was established and it was your turn to use the service (wait times could be up to an hour in some areas).

Obviously this technology was not scalable, and to satisfy the larger consumer market, the radio technology had to change. That change came about with cellular technology.

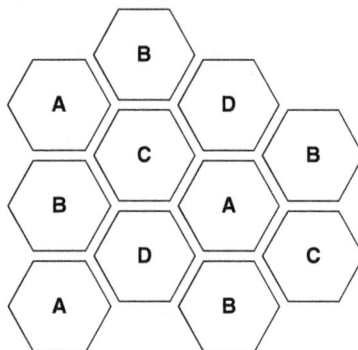

Figure 2.1   Cellular networks reuse frequencies but notice the same frequency can never be used in an adjacent cell.

Radios were coupled with antennas that covered smaller areas. Multiple frequencies were used, so each cell was operating on a different frequency. The frequencies were reused in a geographical area, but they could not overlap one another. The reuse of frequencies in other cells supported more subscribers, and the lower transmission power meant the devices could be smaller, even portable. Cells are not perfectly shaped as shown in Fig. 2.1 but the concept is the same.

When cellular networks reached 2G speeds, the need for data support caused AT&T and Nortel to start a partnership that became known as the Third Generation Partnership Project (3GPP). The partnership was formed between service provider and vendor to define the evolution to an all-IP network.

Others took notice, and soon 3GPP became global. The first task for 3GPP was to define Generic Packet Radio Services (GPRS) and Universal Mobile Telecommunications System (UMTS). This would become known as 3G and resulted in an overlay network that offloaded data traffic from the mobile switching center (MSC) to an IP network dedicated to data.

The standards were published by, European Telecommunications Standards Institute (ETSI), the European standards body. At the same time, the International Telecommunications Union (ITU) was publishing its own standards for Global Services Mobile (GSM) and the air interface, but it was 3GPP that created the standards for the architecture, services, and functions of the network.

While the 3GPP was focused on defining 3G, it didn't stop there; it has continued by defining 4G, and now 5G network standards. These standards include definitions of the network architectures that will support IP from radio tower to network core, and the protocols used within the network.

The Long Term Evolution (LTE) standards introduced with 4G rely on only IP in the network transport, totally replacing the older TDM technology. The architecture also changes to a much flatter, consolidated network.

Ironically, the Diameter protocol did not come from 3GPP, but rather from the Internet Engineering Task Force (IETF). 3GPP has added much to the Diameter standards, building from the base protocol standard by adding new applications for Diameter in both fixed and wireless networks. While early Diameter networks were focused on charging, Diameter today encompasses all control aspects of the 4G networks, and is expanding into 3G, fixed line, and cable networks. Diameter is the most important technology to be developed for telecommunications since the advent of SS7.

GSM Network Architecture

In 0G networks, transceivers were connected to radio towers that served large metropolitan areas. There was no support for roaming, and the devices were far from portable. Once cellular networks made their way into the landscape, devices became smaller thanks to the smaller coverage areas of the cells (and the lower power requirements). 2G networks provided simple voice services. Data consists of email and text messaging, and is not very fast. Browsing the Internet is impossible on these networks.

The radio access network (RAN) in 2G/3G networks consists of two main components. The cell tower and its associated antennas are connected to transceivers in the base transceiver station (BTS). These are typically found in cabinets at the base of the tower. It is common to have numerous radios connecting to the various antennas on the tower, depending on the density of the cell and the types of antenna coverage being supported (Fig. 2.2).

The BTS are connected back to a radio controller, known as the base station controller (BSC). Several BTS are connected to a BSC, which manages the handoffs between radios with the MSC. The BTS connects back to the core network and the MSC. Collectively, the BTS and the BSC comprise a base station subsystem (BSS).

The MSC is responsible for managing handoffs in circuit switched networks when devices move from cell to cell, network to network. The MSC can also provide conversion from circuit-switched to Voice over IP (VoIP) protocols depending on the vendor. No packet data is passed through the MSC, other than simple email and text messaging. The packet core network in 3G is shown in Fig. 2.3.

The home location register (HLR) maintains profiles of the subscribers in the network. Subscribers are assigned to an HLR in their home area (where they make the most calls from, or where they live typically). Included in the subscriber profile are their authentication vectors (used when encrypting over the air interface), the subscriber identity [Mobile Station International Subscriber Directory Number (MSISDN) and International Mobile Subscriber Identity (IMSI)], and the MSC providing the subscriber service (dynamic, as this changes as you roam).

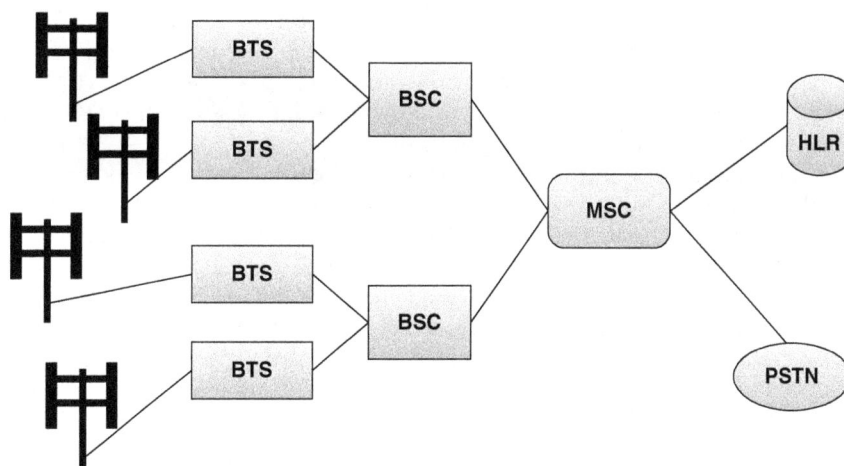

Figure 2.2 A typical GSM 2G network.

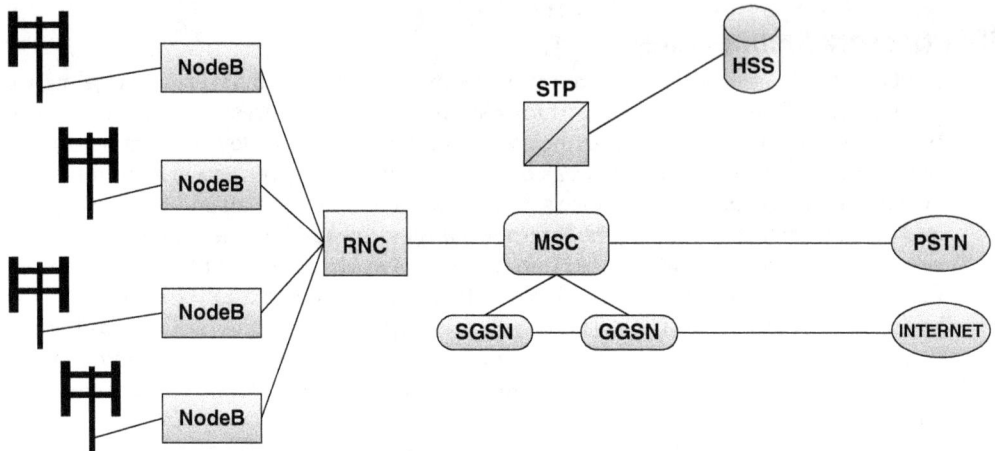

FIGURE 2.3 Simplified view of the GSM 3G network.

The HLR does not maintain accurate location data, as this is not necessary. It knows the serving MSC/VLR for a subscriber, which is enough for routing calls to the device. However, the visitor location register (VLR) does maintain the physical location of the device so it can effectively page the device and connect calls to the device. The VLR and the HLR communicate with each other using SS7 mobile application part (MAP) messages.

Also collocated or integrated with the HLR is the authentication center (AuC). The AuC creates triplets used by the subscriber identity module (SIM) card when the device is activated. This is used as part of the authentication process.

The VLR is often collocated or integrated with the MSC, and is a dynamic database. It maintains subscriber profiles for all active subscribers in the network, even if they are roaming from another network. Authentication vectors, the MSISDN/IMSI, services they are allowed, and the home HLR are maintained in the VLR.

The VLR communicates with other networks so it can authenticate roaming subscribers. For example, when a subscriber roams into another network and requests service, the VLR will query their home HLR to obtain their profile.

Another database maintained in the core is the equipment identity register (EIR). This database contains the international mobile equipment identifiers (IMEIs) of stolen devices. Prior to connecting a call or allowing a data session, the MSC will query the EIR via SS7 to verify the device is not stolen.

EIRs are of local significance, unless the service provider connects with the GSMA IMEI database. This allows service providers around the world to share their stolen device data, and creates a powerful tool for device theft prevention. Sometimes the EIR can be integrated with the HLR, but many times it is a standalone node.

Data in 3G networks are offloaded from the BSS to packet nodes. Data packets being transported within the network are routed through the serving GPRS support node (SGSN). For external data connections, packets are routed to the gateway GPRS support node (GGSN). Think of the SGSN as a large router in the data packet core. Data is offloaded to the SGSN from the BSS, and data encapsulated in the GPRS tunneling protocol (GTP). This eliminates data from being processed at the MSC and wasting processing at the MSC.

The SGSN is responsible for routing of all data packets received, mobility management, authentication of data connections/session, charging of data sessions, and location data from the VLR. The SGSN is similar to the MSC but limited to data only.

The SGSN is responsible for maintaining location of a mobile device whenever data services are initiated. This is analogous with the MSC and the VLR. The SGSN provides VLR capabilities while the mobile is connected to the packet network.

The GGSN acts as the packet data gateway into the network. It hides the packet network from the rest of the world. When connecting to the Internet, it may also convert from the GTP protocol to packet data protocol or other formats as needed. The GGSN also must support charging and authentication functions at the network edge. Figure 2.3 provides a simplified view of 3G networks. Several functions have been left out for clarity.

Diameter/LTE Architecture

LTE introduced a major change to the overall GSM architecture. Network elements were consolidated, and only support IP transport. Everything from the radio to the packet core runs over IP transport. The protocols used in the network were changed as well, with SS7 being replaced by Diameter (for authentication, authorization, and accounting) and the Session Initiation Protocol (SIP).

SIP provides the signaling for voice and multimedia in the network, replacing SS7 ISUP call control. However, 3GPP decided against simple VoIP as the means of supporting voice and standardized on IP Multimedia Subsystem (IMS) as the architecture for the SIP network. Voice over LTE (VoLTE) requires IMS to support voice in 4G networks.

Evolution of the Radio Access Network

The RAN has been consolidated to one network function. The tower connects to the eNode B, which provides the functions of both the BTS and the BSC. Data and voice are both transported by the eNode B to the Mobility Management Entity (MME).

The MME replaces the MSC, and is responsible for managing handoffs when devices are roaming. It also assigns the serving gateway for data routing. Like the MSC, the MME is responsible for authentication when devices request service.

The Evolved Packet Core

When connecting with other networks, the MME becomes a termination point. Interconnections are typically supported through a dedicated gateway MME, to prevent direct routing into the network core. The gateway MME can then hide the network topology from the rest of the world.

The HLR is replaced by a new function that combines the authentication, authorization, accounting (AAA) and HLR functions into one Home Subscriber Server (HSS). The HSS is responsible for the authentication of a subscriber, and authorization for services. Every subscriber is provisioned in the HSS of their home network, and signaling is used to connect to the HSS from the various network elements.

Another addition to the packet core is the policy and charging rules function (PCRF). This new network function allows operators to control the access a subscriber is allowed, based on their device type, subscription plan, and many other factors. Policy is one of the biggest consumers of Diameter bandwidth, because of the number of Diameter transactions created to support data sessions. The PCRF is the function that makes data plans and quota management possible in the 4G network.

According to Oracle's annual LTE Diameter Signaling Index, it is policy driving the most Diameter traffic, especially in networks supporting prepaid charging. The policy server communicates with the charging system throughout a prepaid session, managing the allotted quota for a session, and even managing the actions to be taken by the enforcement points (such as SGSN and PDN Gateway) when the allotted quota has been exhausted.

It is through the packet network that all user data is transported to its destination. The packet network offloads data from the voice switching part of the network, and interfaces with the Internet and other packet networks. With VoLTE, voice is considered packet data as well, and is sent through the packet network. This is a departure from 3G networks where the voice is sent through the circuit switched network.

The packet core also includes the SGSN and the PDN Gateway. These are analogous to the SGSN and GGSN functions in the 3G network. Also note the addition of the Diameter signaling controller (DSC). This function works much like a signal transfer point (STP) in SS7 networks, with many more features and functions. The DSC has become a critical component in the packet core, providing an interconnection point for all Diameter nodes, as well as a gateway into the Diameter network.

The IMS is used for voice sessions. The packetized voice is routed through the IMS that provides additional security and authentication not found in traditional VoIP networks. The IMS also must support Diameter (as well as SIP for the voice signaling). Figure 2.4 provides a very simple (and not totally accurate) view of a 4G network and its components. There are numerous functions omitted for simplicity, and the many interfaces have been reduced to single connection points for clarity.

Evolution of the Signaling Network

Apple changed the telecommunications industry forever when it introduced the iPhone, and soon every device manufacturer was designing smartphones for use in 4G networks. Screens got larger, enticing subscribers to use them as miniature computers. They began using their phones for every aspect of their lives; banking, shopping, social media, and even controlling their homes.

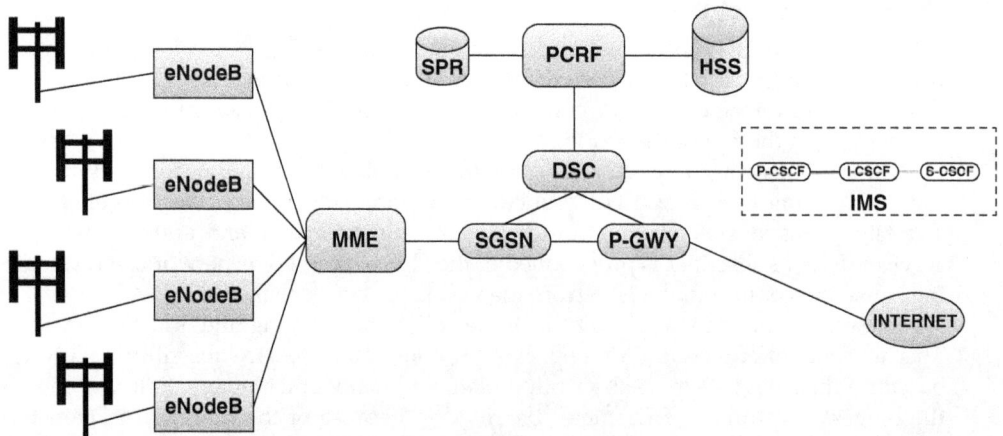

FIGURE 2.4 Over simplistic view of 4G LTE network architecture.

These devices needed 4G speeds to please the user, and policy management gave service providers more opportunity to make money from data connections. But many networks will not make the leap to 4G, electing instead to maintain 3G architectures, or even 2G in some cases.

This decision is determined by economics. In some developing nations, subscribers cannot afford expensive smartphones and their associated data plans, and really don't have a need for their sophisticated applications, therefore there is no market for 4G and beyond. In many networks, 3G and 4G are maintained together, with 4G dominating the metropolitan areas with denser subscriber populations.

Today, it is the adoption of smartphones by the consumer that is driving the growth of Diameter traffic in modern day networks. These types of data connections (such as video streaming and VoLTE) place large demands on the Diameter network and its many interfaces.

The signaling network changes dramatically from 3G to 4G. In a 3G network, SS7 provides communications between the RAN and the network core. The network core is made up of the HLR and MSC.

The MAP protocol is used to support queries to the HLR and the VLR. This supports mobility (moving from one cell to another) and roaming (moving between networks). The MAP protocol also supports queries to the EIR, and the Mobile Location Center (MLC).

The SS7 MAP protocol is used with the customized applications for mobile network enhanced logic (CAMEL) protocol to support services subscribers typically enjoy in their home network while they are outside the network (roaming). For example, if a subscriber needs to change the forwarding on the their phone, the networks would use CAMEL transported through SS7 SCCP to access the HLR/VLR and implement the changes in the subscriber profile.

The messages are universal, regardless of the services being accessed. The MAP messages are somewhat tailored based on the application, but overall the protocol runs the same throughout the network. As network services become more complex, and the ecosystem changes to support revenues from new sources, this is no longer acceptable.

With the introduction of LTE networks and the Diameter protocol, the actions of the protocol change dependent on the application and interface. For example, the HSS is connected to the MME over the S6a interface. The IETF RFC 6733 defines the base protocol to be used for establishing connections over this interface, and managing the sessions. However, it is a different series of 3GPP standards that define the commands and processes to be used for managing Diameter traffic between the MME and the HSS.

These commands may only be used on the S6a interface, and may not be used anywhere else in the network. Included in the commands are hundreds of attribute value pairs (AVPs). Think of these as parameters for each command, containing the actual user data being delivered. When you are talking about more than 50 defined interfaces, you can see where Diameter can become a complex protocol to learn.

In 3G networks, authentication and authorization is performed through the authentication, authorization, accounting (AAA) function, which can be part of the HLR or it can be standalone. This server works much like the HLR does for voice services, but is dedicated to packet network access using the Remote Access Dial-In User Service (RADIUS). This changes in 4G as RADIUS is replaced by Diameter, and the AAA functions are provided at the HSS.

Circuit Switched Network

The circuit switched network supports voice transmission in 2G and 3G networks. The RAN connects to MSCs that act as anchors for the voice calls, and manage connections when subscribers are roaming. MSCs support non-IP facilities such as TDM commonly used in telecommunications networks for voice services. As the network evolves to all-IP, there no longer exists the need for circuit switched facilities.

IMS Network Architecture

The IMS is a packet network supporting voice and video communications. In 4G/LTE, IMS is used to support VoLTE, but was defined many years ago as the next generation packet network for voice services. Think of IMS as VoIP on steroids. Some networks provide VoIP services without using an IMS, but IMS adds another layer of security that is not found in VoIP.

IMS uses SIP and Diameter for signaling. Diameter is used throughout the IMS for authentication and authorization, as well as accounting, but it should be noted that SIP also plays a role in the authentication of a subscriber through the registration process. This book will not go into details about registration as it is outside the scope of this book.

Once the registration process is complete, the HSS is queried to authenticate the subscriber again, this time for network access. Diameter is used to query the HSS in this case. So IMS adds additional components in the network for the purpose of supporting voice and video communications. In 4G LTE networks, IMS is required to support VoLTE.

Mobile Network Addressing

Addressing in the mobile network is critical for things to work. There are many different layers of addressing depending on the service being used, and many other factors. The device is identified, the subscriber is identified, and the network must be identified to support functions such as authentication, authorization, and billing.

Various network nodes use addressing to enable routing. Billing uses various addressing to determine how to charge for services used. All of these addresses are critical, and if misused can allow unauthorized access to the network and its services. For this reason, addresses should be carefully protected through encryption whenever they are used, as well as using other mechanisms such as topology hiding.

International Mobile Subscriber Identity

The IMSI is used to identify a subscriber for authentication and authorization in the network, and is stored in the subscriber identity module (SIM) of a device. Given the sensitivity of an IMSI, networks will also use a temporary IMSI (TMSI) when a mobile connects to a visited network to prevent fraud.

The IMSI is comprised of several identifiers. The first three digits are the mobile country code (MCC). The MCC is followed by the mobile network code (MNC), which can be three digits (North America) or two digits (rest of the world). The MCC and the MNC are defined in ITU E.212 for all networks and countries. The MCC and the MNC combined form a PLMN identifier.

The last part of the IMSI is the MSIN, which is the unique identifier for the subscription.

Mobile Station International Subscriber Directory Number

Simply put, the MSISDN is the telephone number one dials to reach a subscriber, or device. It is assigned to the SIM card of the device, and takes the form of a telephone number in ITU E.164 format. The MSISDN is comprised of the country code, national destination code, and the subscriber number.

In some cases, the MSISDN may be defined as the country code, number planning area (NPA), and subscriber number depending on which standard you read. Regardless, they all have relatively the same definition and it works in the same way.

Mobile Station Roaming Number

The MSRN is a temporary number that gets assigned to a mobile device when it is roaming. Think of this as a routing number used by the home network for forwarding calls to the device when it is roaming in another network.

The MSRN is assigned in the VLR, and sent to the home HLR so calls can be routed to the device when it is in another network. The MSRN is also used in number portability as the routing number for devices that have switched to another service provider. The MSRN is defined in ITU E.164.

Base Station Identification Code

The BSIC is a number that is assigned to each base station. In any network, every network element must have its own address, and the BSIC is the address assigned to the base station. The BSIC is transmitted on the broadcast channel so mobile devices can identify the base stations they are communicating with.

The BSIC is comprised of the network color code (NCC) and the base station color code (BCC), both 3-digit numbers. The NCC identifies the service provider so that if a device is receiving a broadcast signal from two different cell sites, operated by two different service providers, the device can differentiate between the two networks. The BSIC is defined in 3GPP TS 03.03 "Numbering, Addressing and Identification."

Cell ID

The cell ID is a unique number assigned to a base transceiver station (BTS). It can also represent a sector supporting a location area code (LAC) if in a non-GSM network (such as CDMA). There are also some variations in UMTS networks where the number is a combination of the cell ID and the radio network controller (RNC) ID. The cell ID is typically used to identify the cell site serving a specific mobile device, and therefore critical to location based services (as well as for routing to a mobile device).

Regional Subscription Zone Identity

A network uses Regional Subscription Zone Identity (RSZI) to identify regions within its network where roaming is allowed. The RSZI is comprised of the country code, the network destination code (NDC), and the zone code. The zone code is assigned by the network operator while the country code and NDC are defined in ITU E.164.

Location Number

The location number is comprised of the country code, NDC, and a unique location code also called a local significant part (LSP). The LSP is created by the service provider in such a way that it aligns with the country regulators and cannot be mistaken as a MSISDN. The purpose is to identify locations within a network (somewhat analogous to office codes in legacy numbering plans) for routing purposes, as well as to support location-based services.

Service Area Identifier

The Service Area Identifier (SAI) is comprised of the PLMN identifier (the MCC and the MNC combined is the PLMN identifier), the LAC, and the service area code (SAC). The service provider defines the SAC through provisioning. The SAI is then used by the network for defining an area of one or more cells for locating mobile devices in the network.

SS7 Signaling Point Codes

A point code is the address of an SS7 network element. All network elements in the SS7 network must be assigned a point code for routing. The point code format depends on the region. For ANSI networks, the point code is a 24-bit number. The first 8 bits identify the network, the next 8 bits identify a cluster within a network, and the last 8 bits identify the unique network node.

In ITU networks, the format is a 14-bit address with the first 3 bits representing the zone or country, the next 8 bits represent the network, and the last 3 bits represent the network node. ITU networks also must identify International point codes and national point codes to prevent duplication of addresses.

In SS7 routing, the addresses are defined in the routing header as the origination point code and the destination point code. These point codes are used for routing through the SS7 network.

GSN Addresses

Addressing for GPRS nodes follows a specific format in the domain name server (DNS). For example, to identify an SGSN in a network in the US, the logical address might be something like SGSN 1B32, MCC 310, MNC 3. The "SGSN 1B32" is a hexadecimal number assigned by the service provider for the SGSN. The MCC and MNC of course are assigned by the ITU.

There are many more addresses used throughout the network but these are the addresses that one will most likely come across when dealing with Diameter protocol messages. For a complete description of addresses used in GSM networks, go to 3GPP TS 23.003.

Access Point Names

The Access Point Name (APN) identifies the data network that a subscriber wants to connect with. For example, if the subscriber is trying to connect to the Internet in the Claro Dominicana network, the APN would be internet.ideasclaro.com.do.

The APN has two parts. The network identifier defines the external network that the GPRS is connected to. The operator identifier defines the operators' packet domain network. The operator identifier uses the MCC and the MNC to uniquely identify service provider networks. While the network identifier is mandatory, the operator identifier is optional.

CHAPTER 3

Introduction to Diameter

Why Diameter

The Remote Authentication Dial-In User Service (RADIUS) protocol was originally developed to support Point-to-Point Protocol (PPP) connections. RADIUS managed authentication, authorization, and accounting over these dial-up connections. Networks have evolved a great deal over the years, and RADIUS quickly became too limited for modern-day services, especially in today's wireless networks.

New demands from wireless devices connecting to the data network in wireless networks were beyond the capabilities of RADIUS, and the Internet Engineering Task Force (IETF) knew it had to evolve the RADIUS protocol. The answer is Diameter.

The IETF has developed the base set of requirements for Diameter. This base protocol is defined in RFC 6733, and must be supported by all Diameter nodes at a minimum. There was an earlier version (RFC 3588) that was later replaced with RFC 6733, but you will still find many references to the older RFC. They are both backward compatible. This ensures that all nodes are able to communicate with one another via a standards based network connection.

Diameter was designed as an extensible protocol. This means the protocol can continue to evolve, without having to affect all the existing network elements. The base protocol ensures that all network elements can continue to communicate as defined. This is accomplished by extending the protocol one of two ways:

- Adding a new command to an application
- Adding a new attribute value pair (AVP) to an existing command
- Creating an entirely new application with its own set of commands and AVPs

New applications introduced into the network are defined as an interface in the Diameter network. That interface supports a specific set of messages (commands) and user data is delivered via attribute value pairs (AVPs).

For example, in the base protocol, AVPs are used to deliver authentication data (such as user credentials), routing information, and even usage information for charging. The applications are the primary users of this data.

The Third Generation Partnership Project (3GPP) has been busy defining new applications and interfaces for the Diameter network. While they are supporting the

15

base protocol defined in RFC 6733, there may be some AVPs that 3GPP has chosen not to support for specific applications. Therefore, you have to treat each interface differently, and use the specifications for that specific interface. In this book, you will find both the IETF defined AVPs for base commands, as well as 3GPP amendments and modifications defined for 3GPP wireless networks.

Vendors can develop their own proprietary commands and AVPs, as long as they also support the base protocol as defined. When a network element receives a Diameter request, if there are AVPs within the request it does not understand, it simply ignores the portions it cannot translate, and processes the AVPs it does understand. The only requirement is that the AVPs must conform to the protocol structure as defined in RFC 6733. If changes are to be made to the base protocol, the IETF must approve those changes and introduce the change as a new version of the Diameter protocol specification.

The base protocol can be used on its own to support accounting in the network. The original implementation of Diameter was developed for charging networks and the replacement of the RADIUS protocol in charging networks. However, 3GPP has added additional requirements for charging in a wireless domain, so if the application is to support billing in wireless networks, RFC 6733 is not enough, and the 3GPP specifications must be followed.

Diameter versus RADIUS

While RADIUS is an acronym, Diameter is not. The name comes about as an engineering joke. The diameter of a surface is twice its radius; therefore, since Diameter is twice what the RADIUS protocol could provide, Diameter is twice the RADIUS.

There are many different shortcomings in the RADIUS protocol. Remember that this protocol was designed to support dial-up services. Today's networks are packet based, using IP as the transport, and the services that are supported are far more complex than earlier network services.

Failover

RADIUS was not able to provide fail-over procedures. There is no means for servers to communicate that they are going out of service, or for the orderly termination of sessions for any reason. When an error occurs in RADIUS, there are no procedures for trying to rectify the error. The session simply fails. Part of supporting a failover process is maintaining the state if sessions.

Diameter supports both stateful and stateless sessions, whereas RADIUS does not support stateful sessions. This means nodes in the RADIUS network do not have any knowledge of the state of a session.

Transmission-Level Security

RADIUS assumes that security is managed in the back office billing systems rather than in the network. This stems from the notion that network connections between service providers can be trusted. Nothing could be further from the truth today.

In today's world, security must be implemented at all levels, using layered security architecture. Started at the transmission level, all the way up to the application level.

One of the issues in Signaling System #7 (SS7) networks is the lack of authentication at the transport layer (between networks). This is resolved through the use of IPsec. (Note that IETF originally required TLS/DTLS encryption, but later recognized IPsec as an acceptable alternative).

Diameter supports the use of encryption at the transport layer. This is important in today's implementations where we are seeing Diameter replacing SS7 Mobile Application Part (MAP) applications in the 4G/LTE wireless network.

Reliable Transport
RADIUS uses the User Data Protocol (UDP) as its transport, a very unreliable Internet transport. In today's sophisticated wireless networks, reliable transport is paramount to supporting real-time applications such as video and real-time gaming.

In Diameter networks, the more reliable Stream Control Transmission Protocol (SCTP) is used. SCTP was developed to support many real-time applications in the telecommunications domain; originally for SIGTRAN (SS7 over IP).

There are many procedures used during the connection phase to ensure a reliable connection, as well as methods for dealing with transport failures and protocol errors. These will be explained in this chapter later on.

Agent Support
The RADIUS protocol assumes that every connection is a direct connection, without using agents in the transport stream. This does not allow proxies or redirect agents to be used to route messages through the network. This was fine for a dialup connection where connections were made with routers in the data domain, but today wireless networks need the ability to use proxies to determine the best routing for specific message types.

Server-Initiated Messages
Diameter is considered a Peer-to-Peer Protocol. The client generates a Diameter request and sends to an application acting as the server. The Diameter server is then responsible for processing the request and sending an answer. RADIUS does not use this concept.

The request is sent through the network using Diameter agents. The agents define how the request will be routed to its destination. The use of agents allows for functions that can manage complex routing schemes and session binding in the network. This is an important capability of Diameter routers.

Transition Support
Networks are evolving to Diameter, but not all at once. Most networks are migrating portions of the network to IP, and hence Diameter for signaling. The "legacy" portion of the network is maintained for cost reasons, until it becomes necessary (and cost effective) to replace the equipment in that portion of the network. The 3GPP is defining new Diameter interfaces to support even portions of the legacy network, allowing service providers to upgrade their older platforms to support Diameter without migrating the whole network to LTE.

Each connection to an application is defined by an interface, which has its own set of procedures and Diameter messages. This allows networks to transition to Diameter while still maintaining their legacy protocols. Interworking is used between technology types to convert between Diameter and SS7, as an example.

As applications evolve and new connection points are implemented, network elements supporting the base protocol can be easily added into the network, even though they may be supporting a new application not defined by the IETF.

The 3GPP continues to define new applications for Diameter, and new interfaces for the Diameter network. These interfaces support the base protocol, as well as an application specific interface with its own set of commands and AVPs.

Capability Negotiation

One of the shortcomings of the RADIUS protocol is the lack of negotiation between network elements during the connection phase. This means incompatible sessions fail. The Diameter protocol solves this by offering capabilities negotiation prior to establishing a connection.

Prior to a session being established, both entities must exchange their capabilities to ensure compatibility between the two network functions. Error handling manages any incompatibility issues during the exchange.

Peer Discovery and Negotiation   RADIUS requires manual configuration and provisioning, while Diameter supports peer discovery through the Domain Name Service (DNS). However, this author for security reasons does not recommend peer discovery. Static IP addressing is more secure, even though it requires additional provisioning in the network elements.

The concept of peer discovery is to allow new nodes to be added to the network autonomously, as the network detects them. While this can be used in some parts of the network, it presents many problems when used at network interconnects.

Roaming Support   Perhaps the most important aspect of Diameter is the ability to support sessions while the device is roaming between networks. RADIUS does not support any concept of roaming, but Diameter must be able to support users and their devices regardless of location.

Previous to 4G/LTE, networks used the SS7 MAP protocol to support authentication and authorization in the network while roaming. Because of the ability for Diameter to support agents, the protocol can be used to replace the MAP protocol in IP networks such as LTE.

This also consolidates the number of protocols required in wireless networks. Rather than having to support RADIUS and SS7, networks only need to support Diameter for authentication and authorization, as well as roaming. Many more applications in the wireless network are being defined for Diameter support.

Diameter Base Protocol

The Diameter base protocol as defined by the IETF supports establishment of a connection, transport of Diameter messages, and supporting accounting applications. RFC 6733 is the latest requirements document from the IETF as of the writing of this book.

The base protocol could be used by organizations using Diameter for accounting networks, but in wireless networks, the 3GPP has added additional AVPs to the base protocol for accounting services. They will be reflected here in this book.

Basic Message Structure

The base protocol is defined in IETF RFC 6733 (version 1). The base protocol defines the minimal set of requirements that must be met to be able to route messages in a Diameter network, and includes processing requirements for accounting applications. The 3GPP has defined additional applications for Diameter as used in 3GPP networks, and should be considered as an accompaniment depending on the application to be supported.

0 1 2 3 4 5 6 7 8 9 0 1 2 3 4 5 6 7 8 9 0 1 2 3 4 5 6 7 8 9 0 1

Version	Message Length		
Command Flags	Command Code		
Application ID			
Hop-by-Hop Identifier			
End-to-End Identifier			
AVPs			

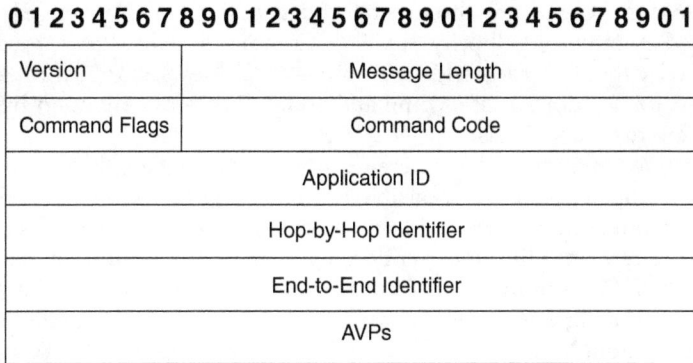

FIGURE 3.1 Basic message structure.

For example, if designing an application such as the Home Subscriber Server (HSS), the node will have to support the base protocol, in addition to the 3GPP requirements for the S6a interface. The two therefore work together.

The basic structure of the Diameter protocol is shown in Fig. 3.1.

The first field is the protocol version. As of this book, the version is one (1). Any changes to the protocol would require a new RFC to be published defining the changes. All changes to the protocol must be backward compatible, to prevent breaking of the network.

The message length is always in multiples of four. The length includes the command header (shown in Fig. 3.1) and all padded AVPs. To fill in space where the message cannot fit a multiple of four, the remaining space is filled with zeroes for padding.

There are eight 1-bit command flags. These flags are defined by the IETF and cannot be changed or amended without ratification by the IETF community. The first flag is the "R" bit, indicating the message is a request (if equals one) or an answer (if equals 0).

The "P" bit with a value of one indicates a message that can be proxied, relayed, or redirected. If the "P" bit is equal to zero, than the message must be processed by the receiving node (if capable). The receiving node can reject the message if it cannot process it. Error processing would be used to send an answer message back to the originator, with the appropriate reason code.

The "E" bit is the error bit, and when set to a value of one (1), indicates an error occurred. If a request is received with an error detected, the receiving node will create an answer message and return to the originator of the request. The answer message will contain the error information (usually in the Result-Code AVP) in place of the normal answer message. The error bit is only used in an answer message, and never sent in a request message.

The "T bit is used to indicate a potentially retransmitted message. If a request is sent but no answer is received, the server will resend the request with the "T" bit set to a value of one (1). This notifies the receiving node that this message could have been received already, and the subsequent message is a retransmit of the original request.

The remaining bits are reserved, and therefore will be set to a value of zero. They are reserved for future revisions of the protocol, and if one of the reserved bits is received with a value other than zero (0), will create a protocol error.

Command codes are three octet codes defined and managed by the Internet Assigned Numbers Authority (IANA). Command codes 256 through 8,388,607 are reserved by the IETF for standard commands. Values 8,388,608 through 16,777,213 are reserved for vendor specific command codes. These are allocated by the IANA on a first-come-first-serve basis.

A vendor must provide a reference specification where the command code will be used, or a contact person responsible for managing the command code. This prevents vendors from simply reserving large blocks of codes that will never be used. It also provides a reference for other vendors wishing to process the command if applicable. Values 16,777,214 through 16,277,215 are reserved for experimental use.

Every command is defined along with its code, which is found in the header of the message. Throughout this book, we will be referencing both the command name and the command code.

The application ID is used to identify the application the request is for. This is a four-octet field that identifies authentication, accounting, or vendor specific applications.

The hop-by-hop identifier is used to help match a request to an answer. Each node is responsible for sending its own unique identifier when forwarding a message, and when it receives an answer, it must use the same identifier received to return to the originator. The node for correlation of messages locally uses it.

The end-to-end identifier is used from the host (originator) to the destination. The originator sets the identifier and it is passed as is to the destination. When an answer is returned, it must contain the same end-to-end identifier as the request so the originator can correlate the answer to the request. This is also used for finding duplicate messages by looking at this and the Origin-Host AVP. The value of the end-to-end identifier is based on the local time (first 12 bits) with the remaining 20 bits generated randomly.

Commands and Command Codes

The Diameter protocol is based on commands. Each command consists of a request and an answer. Together, the two messages constitute a transaction. Both the request and the answer will contain user data, in the form of attribute value pairs (AVPs). Both the request and the answer will have the same code. Note the use of the "R" bit above, where the request is identified in the header.

When a request is sent, the request must be processed prior to sending the answer. The answer message will identify if the request was successful or not, and may provide additional data needed to complete the request. If the receiver of a request is not able to complete the request, it may redirect the request to another node for processing, which will then send its own answer when the processing has been completed.

Each interface (and therefore each application) will have its own set of commands and usually AVPs. Many AVPs can be shared across multiple interfaces, while some are dedicated to specific applications and found only on those interfaces. The same is true of commands. Some commands can be shared across different interfaces, but there are many that are only found on one interface.

Rather than duplicating the message header and the other common portions of the Diameter messages, we will abbreviate the headers. In this book, we will use the notation found in the standards for consistency. The following are the conventions used for showing Diameter commands and their AVPs.

The beginning of each of the messages will usually start with the message acronym in brackets, followed by two colons and the equal sign. Then another set of brackets representing the Diameter header, the command or AVP code, and an indicator if the message is required and can be proxied. This is the same format used in the RFC and 3GPP specs, so I am using it here as well. Below is an example:

<ASR> :: = <Diameter HDR: 274, REQ, PXY>

This will then be followed by the AVPs used in the command (or if showing a grouped AVP, the AVPs used within the group). For each of the AVPs represented, the following notation is used.

< > indicates the AVP is in a fixed position, and mandatory

{ } indicates an AVP is mandatory, but can be in any position within the message

[ ] indicates optional AVPs. They can be positioned anywhere in the message

* indicates minimum/maximum values if an optional AVP requires these values and precedes it 1 * indicates the number of times the element is present in the message (the AVP can appear multiple times).

The content of AVPs is formatted in a number of different ways. The following defines the formatting used in Diameter AVP content:

- Grouped—a grouped AVP contains additional AVPs, which can also be grouped or can be used outside of a grouped AVP (often referred to as command level).
- OctetString—this is a string of binary data with a length in multiples of 8 (one octet).
- Enumerated—this is variable data with at least 3 states according to ASN.1. In Diameter, you will find this content to be represented by a number, with the number representing some predefined value according to the standards.
- Unsigned32—the value is formatted using long integers (32 bits).
- UTF8String—UTF8 is the most used character encoding, and is backward compatible with ASCII encoding.
- DiameterURI—the DiameterURI contains the fully qualified domain name (FQDN), port, transport (SCTP), and protocol (always Diameter in this case).

These formats are identified in each of the chapters for the relevant AVPs.

Attribute Value Pairs

User data is sent in the form of attribute value pairs (AVPs) sent in the command request and answer. The term is a carryover from database terminology. There are hundreds of defined AVPs to date, and the list is growing longer. Each time a new application is defined, new AVPs are defined to request data and return data from the application.

Basic AVP Formats As mentioned earlier, the AVP contains the user data being sent to an application, or being returned from an application. AVPs must follow a specific format, even if they are vendor specific. The basic format is shown in Fig. 3.2.

The AVP code, like the command code, is unique for each AVP. Codes 1-255 are reserved for cases where RADIUS attributes are being reused. All codes 256 and above

```
0 1 2 3 4 5 6 7 8 9 0 1 2 3 4 5 6 7 8 9 0 1 2 3 4 5 6 7 8 9 0 1
```

AVP Code	
VMPrrrr	AVP Length
Vendor ID (opt)	
Data......	

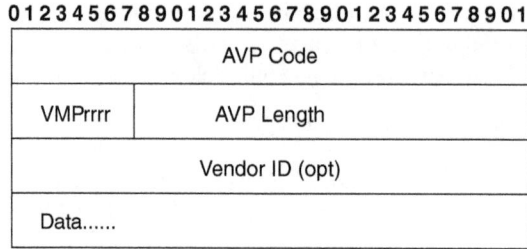

FIGURE 3.2 Basic AVP structure.

are defined by the IANA for use in Diameter. Vendors submit their AVPs to the IANA for number assignment.

The "V" bit is the vendor bit, and will be set to the value of one when a vendor ID is present. Vendors can define their own proprietary AVPs, which is what makes Diameter such a powerful protocol. It is extensible, allowing for easy additions to be made to the protocol to support new applications, without requiring changes to the base protocol.

Vendors with their own ID registered with the IANA will also receive a range of AVP code values. This prevents "collision" with other vendor messages caused by duplicate vendor AVPs. These values are dedicated to the registered vendor, and can be found at www.iana.org/assignments/enterprise-numbers.

The "M" bit is the mandatory bit. This bit tells the recipient that the AVP and its associated data is mandatory (and therefore must be read). If the "M" bit is set to a value of zero, then the AVP can be ignored. Some AVPs might be mandatory on one interface, but when used on another interface, it is not mandatory.

The "P" bit is reserved for future releases to support end-to-end security, which has not been defined at this time. The "P" bit will contain a value of zero at this time. The remaining bits are reserved for future protocol versions and therefore will be set to a value of zero.

The AVP length includes the code field, AVP length, AVP flags, vendor ID, and all user data. Since AVPs are variable in nature, there are a number of options available as indicated in the AVP length field:

- At least 8 = octet string (12 if "V" bit equals 1)
- 12 (16 if "V" bit equals 1) = integer 32
- 16 (20 if "V" bit equals 1) = integer 64
- 12 (16 if "V" bit equals 1) = unsigned 32
- 16 (20 if "V" bit equals 1) = unsigned 64
- 12 (16 if "V" bit equals 1) = float 32
- 16 (20 if "V" bit equals 1) = float 64
- 8 (12 if "V" bit equals 1) = grouped AVP

The Vendor-ID field contains the unique vendor identifier. Vendors who wish to add their own proprietary AVPs to the Diameter protocol must register with the IANA, and receive their Vendor-ID. This field lets receiving nodes know if they should ignore the AVP, or process the AVP. If the AVP is an IETF AVP, the value will always be zero.

AVPs align to a 32-bit boundary. Zeroes are used as padding when the data cannot fill the boundary. Padding is not counted in the message length of AVPs, but will be counted in the command header where the length of the entire message (command and AVPs) is given.

Grouped AVPs   A group of AVPs can be included in the data field of one AVP, forming a grouped AVP. Since all of the AVPs are linked, all AVPs must be received before the grouped AVP can be processed. The AVPs themselves will be listed throughout this book, and will be identified as a grouped AVP.

Base Protocol AVPs   The IETF has defined the base protocol to be supported by all nodes in a Diameter network. This is the minimal set of requirements that must be supported for nodes to be able to communicate. In addition to the base protocol, other requirements can be supported depending on the application (and therefore the interface). The list of base AVPs can be found in Table 3.1.

Attribute Name	Code	Attribute Name	Code
Acct-Interim-Interval	85	Host-IP-Address	257
Accounting-Realtime-Required	483	Inband-Security-ID	299
Acct-Multi-Session-ID	50	Multi-Round-Time-Out	272
Accounting-Record-Number	485	Origin-Host	264
Accounting-Record-Type	480	Origin-Realm	296
Acct-Session-ID	44	Origin-State-ID	278
Accounting-Sub-Session-ID	287	Product-Name	269
Acct-Application-ID	259	Proxy-Host	280
Auth-Application-ID	258	Proxy-Info	284
Auth-Request-Type	274	Proxy-State	33
Authorization-Lifetime	291	Redirect-Host	292
Auth-Grace-Period	276	Redirect-Host-Usage	261
Auth-Session-State	277	Redirect-Max-Cache-Time	262
Re-Auth-Request-Type	285	Result-Code	268
Class	25	Route-Record	282
Destination-Host	293	Session-ID	263
Destination-Realm	283	Session-Timeout	27
Disconnect-Cause	273	Session-Binding	270
Error-Message	281	Session-Server-Failover	271
Error-Reporting-Host	294	Supported-Vendor-ID	265
Event-Timestamp	55	Termination-Cause	295
Experimental-Result	297	User-Name	1
Experimental-Result-Code	298	Vendor-ID	266
Failed-AVP	279	Vendor-Specific-Application-ID	260
Firmware-Revision	267		

TABLE 3.1   Base Diameter AVPs as Defined by the IETF

Base Protocol Commands

The base protocol commands that are defined in the IETF RFC 6733 are used for transport and connection establishment in Diameter networks, and also to support accounting applications. As a bare minimum, any node in the network that wishes to connect to another node via Diameter must support the base protocol as defined in RFC 6733. In this book, we will continue to reference the base protocol, meaning the requirements defined in RFC 6733.

It should be noted that the original RFC 3588 was rewritten and submitted as 6733. Most of the work done in 6733 was cleanup and clarification from the original RFC. The protocol really did not change, as this would have required a new version of Diameter to be declared. Therefore, the industry is still working from Version 1 of the protocol.

This book is based on the RFC 6733, as well as numerous 3GPP standards defining the requirements for various applications. It can be assumed then that the 3GPP requirements are for wireless networks (3G and 4G).

Many have a tendency to equate Diameter only to LTE networks, but Diameter can actually be found in non-LTE networks as well. Remember that Diameter was originally developed to replace the RADIUS protocol, and early implementations were in the data centers supporting charging platforms.

There are many new interfaces being defined for Diameter in 3G networks as well, and at least a few service providers have begun using Diameter in their 3G domains. So don't limit the scope of Diameter to simply LTE as it is much broader than that, and you may find implementations of Diameter that are not 3GPP compliant.

Abort-Session-Request/Answer

The Abort-Session-Request/Answer (ASR/ASA) is used to stop a session in progress. For example, if a session is in progress for a video and the prepaid subscriber runs out of quota, the policy and charging rules function (PCRF) could send an ASR to the PGW to stop the session.

The PGW can then decide if it should stop or ignore the command. Network nodes can be configured to accept ASRs from specific addresses (such as the PCRF) and ignore all others. In the ASA, the node describes what action it took (Table 3.2).

Accounting-Request/Answer 271

The Accounting-Request/Answer (ACR/ACA) is used when a network element wants to send accounting information to the accounting server. The ACA is used as the response to the ACR. The home server only uses this, as the home server is the only server that should respond to an ACR. This makes sense because the accounting is done in the home network.

The items that are lined out are defined in the IETF RFC 6733, but not used by 3GPP networks. Those in bold were added by the 3GPP in TS 32.299 (Table 3.3).

Capabilities-Exchange-Request/Answer 257

The CER/CEA command is used to exchange capabilities between two network elements. The nodes communicate the applications they support so the sending node understands the capabilities of the receiving node, and the AVPs that it will be able to support. The Host-IP-Address AVP identifies each IP address that may be used for sending Diameter messages.

ASR	ASA
<Session-ID>	<Session-ID>
{Origin-Host}	{Result-Code}
{Origin-Realm}	{Origin-Host}
{Destination-Realm}	{Origin-Realm}
{Destination-Host}	[User-Name]
{Auth-Authentication-ID}	[Origin-State-ID]
[User-Name]	[Error-Message]
[Origin-State-ID]	[Error-Reporting-Host]
*[Proxy-Info]	[Failed-AVP]
*[Route-Record]	*[Redirect-Host]
*[AVP]	[Redirect-Host-Usage]
	[Redirect-Max-Cache-Time]
	*[Proxy-Info]
	*[AVP]

TABLE 3.2 ASR/ASA Command Content

ACR	ACA
<Session-ID>	<Session-ID>
{Origin-Host}	{Result-Code}
{Origin-Realm}	{Origin-Host}
{Destination-Host}	{Origin-Realm}
{Destination-Realm}	{Accounting-Record-Type}
{Accounting-Record-Type}	{Accounting-Record-Number}
{Accounting-Record-Number}	[Acct-Application-ID]
[Acct-Application-ID]	[Vendor-Specific-Application-ID]
[Vendor-Specific-Application-ID]	[User-Name]
[User-Name]	[Accounting-Sub-Session-ID]
[Destination-Host]	[Acct-Session-ID]
[Accounting-Sub-Session-ID]	[Acct-Multi-Session-ID]
[Acct-Session-ID]	[Error-Message]
[Acct-Multi-Session-ID]	[Error-Reporting-Host]
[Acct-Interim-Interval]	[Failed-AVP]
[Accounting-Realtime-Required]	[Acct-Interim-Interval]
[Origin-State-ID]	[Accounting-Realtime-Required]
[Event-Timestamp]	[Origin-State-ID]
*[Route-Record]	[Event-Timestamp]
*[Proxy-Info]	*[Proxy-Info]
[Service-Context-ID]	*[AVP]
[Service-Information]	
*[AVP]	

TABLE 3.3 ACR/ACA Command Content

CER	CEA
{Origin-Host}	{Result-Code}
{Origin-Realm}	{Origin-Host}
1*{Host-IP-Address}	{Origin-Realm}
{Vendor-ID}	1*{Host-IP-Address}
{Product-Name}	{Vendor-ID}
[Origin-State-ID]	{Product-Name}
*[Supported-Vendor-ID]	[Origin-State-ID]
*[Auth-Application-ID]	[Error-Message]
*[Inband-Security-ID]	[Failed-AVP]
*[Acct-Application-ID]	*[Supported-Vendor-ID]
*[Vendor-Specific-Application-ID]	*[Auth-Application-ID]
[Firmware-Revision]	*[Inband-Security-ID]
*[AVP]	*[Acct-Application-ID]
	*[Vendor-Specific-Application-ID]
	[Firmware-Revision]
	*[AVP]

TABLE **3.4**   CER/CEA Command Content

Given the sensitivity of the contents in this command, it should never be sent across network boundaries without encryption. This prevents man-in-the-middle attacks from snooping hackers, and mitigates the ability to obtain network sensitive information that can be used for exploiting an interconnect as we have seen in SS7 networks.

The command uses the following format (Table 3.4). Note that there is no Session-ID in this command, because a session has not yet been established.

Device-Watchdog-Request/Answer 280

A Device-Watchdog-Request (DWR) may be sent to an adjacent peer during periods where no traffic is being sent. This is used to verify that the connection is still in place and working properly. If no Device-Watchdog-Answer (DWA) is received, then the sending node assumes that there is a transport failure.

The watchdog message is part of the state machine that is maintained in all Diameter nodes. State is local—there is no knowledge of remote nodes kept in memory. This is the equivalent to SS7 MTP L2, where nodes use link management to verify connections (Table 3.5).

Disconnect-Peer-Request/Answer (DPR/DPA) 282

This is used when a node needs to disconnect from a peer. The Disconnect-Cause AVP identifies the reason the connection is being disconnected. This command prevents nodes from trying to reconnect when connections are closed because of congestion or because the node does not want to maintain a connection (Table 3.6).

DWR	DWA
{Origin-Host}	{Result-Code}
{Origin-Realm}	{Origin-Host}
[Origin-State-ID]	{Origin-Realm}
*[AVP]	[Error-Message]
	[Failed-AVP]
	[Origin-State-ID]
	*[AVP]

TABLE 3.5   DWR/DWA Command Content

DPR	DPA
{Origin-Host}	{Result-Code}
{Origin-Realm}	{Origin-Host}
{Disconnect-Cause}	{Origin-Realm}
*[AVP]	[Error-Message]
	[Failed-AVP]
	*[AVP]

TABLE 3.6   DPR/DPA Command Content

The Disconnect-Cause AVP uses the following values:

0 = REBOOTING

1 = BUSY

2 = DO_NOT_WANT_TO_TALK_TO_YOU

If the value is REBOOTING (0), the node can attempt to reconnect. However, if the value is BUSY (1), the receiving node should not attempt to reconnect as the sender is congested. If the value is DO_NOT_WANT_TO_TALK_TO_YOU (2), the sender does not see a need for this connection and therefore is closing the connection. Reconnection should not be attempted.

Re-Auth-Request/Answer (RAR/RAA) 258
This command is typically used in prepaid accounting (online charging). A server sends this command when it needs to know if a subscriber is still using a service. It is sent to the access node supporting the service to the subscriber, and forces reauthentication/authorization (Table 3.7).

Session-Termination-Request/Answer 275
This command is only used when a server is maintaining state of Diameter sessions. Any time a session is terminated, the Session-Termination-Request (STR) must be sent to the access device providing the service to notify it of the session termination.

RAR	RAA
<Session-ID>	<Session-ID>
{Origin-Host}	{Result-Code}
{Origin-Realm}	{Origin-Host}
{Destination-Realm}	{Origin-Realm}
{Destination-Host}	[User-Name]
{Auth-Application-ID}	[Origin-State-ID]
{Re-Auth-Request-Type}	[Error-Message]
[User-Name]	[Error-Reporting-Host]
[Origin-State-ID]	[Failed-AVP]
*[Proxy-Info]	*[Redirect-Host]
*[Route-Record]	[Redirect-Host-Usage]
*[AVP]	[Redirect-Max-Cache-Time]
	*[Proxy-Info]
	*[AVP]

TABLE 3.7 RAR/RAA Command Content

This also applies if a proxy is unable to support an authorized session (for example, a mandatory parameter is not supported). The proxy sends the STR to notify of the session termination even if no session was created.

The STR is sent as part of the session clearing if the Authorization-Lifetime expires and no Re-Auth-Request is received (Table 3.8).

STR	STA
<Session-ID>	<Session-ID>
{Origin-Host}	{Result-Code}
{Origin-Realm}	{Origin-Host}
{Destination-Realm}	{Origin-Realm}
{Auth-Application-ID}	[User-Name]
{Termination-Cause}	*[Class]
[User-Name]	[Error-Message]
[Destination-Host]	[Error-Reporting-Host]
*[Class]	[Failed-AVP]
[Origin-State-ID]	[Origin-State-ID]
*[Proxy-Info]	*[Redirect-Host]
*[Route-Record]	[Redirect-Host-Usage]
*[AVP]	[Redirect-Max-Cache-Time]
	*[Proxy-Info]
	*[AVP]

TABLE 3.8 STR/STA Command Content

Connecting to Diameter Nodes

A connection is a physical transport-layer connection between two peers. Connections and sessions are independent of one another. A session is a logical dialog between two peers. Sessions are identified by session IDs, and there can be more than one session on a connection.

The Diameter protocol runs on port 3868 for TCP or SCTP connections, and port 5658 if encryption (TLS/DTLS) is being used. TLS is used over TCP while DTLS is used over SCTP connections. Port 3868 can be used with TLS/DTLS, but the capabilities exchange will not be encrypted and will be sent in clear text. This is not ideal, as the data within the capabilities exchange message set contains sensitive user data such as cryptographic keys and credentials.

The first RFC to define Diameter was RFC 3588. This RFC did not support security at the transport level, and was replaced by RFC 6733. The new RFC did not change the version of the protocol, but rather clarified a number of ambiguous aspects of the base protocol and added support of encryption at the transport layer.

Peer Connections

There are specific Diameter messages reserved for the establishment of a connection between peers, and therefore supported by all applications. They are identified by the Application-ID of zero (0). They are defined in the base protocol RFC 6733. Those messages are

Capabilities-Exchange-Request/Answer (CER/CEA)

Device-Watchdog-Request/Answer (DWR/DWA)

Disconnect-Peer-Request/Answer (DPR/DPA)

Diameter connections between peers require TLS/DTLS to prevent user data from being exposed. IPsec is an alternative to using TLS/DTLS, and is recommended by the Groupe Speciale Mobile Association (GSMA) in IR.88. Indeed we are seeing potential in SS7 networks today where the protocol is being used at interconnects to retrieve subscriber data. This is possible because SS7 interconnections do not use encryption, and do not authenticate the connection.

Diameter peers will attempt to connect periodically when there is no activity in a session provided a connection exists. The default is every 30 seconds based on timer T_c. This is configurable in the network elements.

Typically Diameter messages are sent in order, but can be sent out of order provided the network has made provisions for preventing head-of-line blocking. RFC 6733 recommends sending the command Device Watchdog Request (DWR). When this is used, the receiver returns Device Watchdog Answer (DWA), and out-of-order processing can begin.

Peer Discovery

The IETF supports the use of peer discovery in Diameter networks. This allows new network elements to be added to the network without having to provision the new element in routing tables manually. However, peer discovery also introduces a security threat, and therefore not recommended by this author.

Instead, static IP addressing should be used, especially when connecting two networks. This provides an extra layer of confidence that the network being connected incorporates a gateway function and the IP address given is legitimate. Of course, this is not guaranteed, but it should have a higher confidence level than a peer discovered via DNS without any human validation. If a Fully Qualified Domain Name (FQDN) is used for the destination, it will require resolution by the DNS.

The peer must be provisioned in a routing table to be reachable (also known as a peer table). There should be at least two peers defined per realm, a primary and a secondary. This follows the design concepts for earlier signaling methods and prevents message loss during outages. There can be more than just the two peers, but there should be at least two peers per realm. Then load balancing should be used between all the peers for a particular realm.

Capabilities Exchange

During the connection phase, there must be a capabilities exchange. This allows a node to receive a peer's identity and the applications it supports. The identity of the requesting peer is not known until CER is received. This should always be done on an encrypted link to prevent eavesdropping.

TLS/DTLS allows for messages to be sent over multiple interfaces, which requires one Host-IP-Address AVP for each IP address that is used when sending a Diameter message. All of the IP addresses cannot be placed in one AVP, so it is possible you will see multiple Host-IP-Address AVPs in both the CER and the CEA.

Once a connection request has been received, TLS/DTLS handshaking must be completed prior to sending CER. This prevents CER from being sent without encryption, but if encryption is not used, then of course there is no handshake. Given the current climate with interconnects, and the vulnerabilities already being realized, TLS or IPsec should be used prior to CER transmission.

When a peer connection is being made, the requesting peer will request the capabilities exchange. The peers exchange messages containing the application IDs for all applications supported by that peer. This ensures that there are no issues with compatibility in the session. If any other message is received on a connection prior to CER, the message is discarded and the connection closed. Likewise, if a connection is made and CER is not received within a specified period of time, the connection is closed.

The receiver of a CER checks the Application-IDs returned and compares them with its own capabilities. If there are applications the receiver does not support, the receiver will then send a CEA with the Result-Code AVP containing the value DIAMETER_NO_COMMON_APPLICATION. CER and CEA commands should not be proxied, relayed, or redirected.

Relay and proxy agents use the application ID to find servers upstream that support the application being requested, or they will return an answer with the Result-Code AVP set to the value DIAMETER_UNABLE_TO_DELIVER.

Diameter relay and redirect agents assume the sender of the Capabilities Exchange Request (CER) command to support all applications. They will send the application ID with a value of relay agent. Table 3.9 shows the values for application ID.

Diameter common message	0
Diameter base accounting	3
Diameter relay	0xffffffff

TABLE 3.9 Application ID Values

Peer State Machine

A peer state machine is kept for all Diameter peers. Status can be either open or closed for each peer. A peer is not paired in the state machine until a CER is received. It is then the host is identified and its state is updated.

Hop-by-Hop Identifier   The Hop-by-Hop-Identifier is always set to a number unique to the node sending the message, even when forwarding a message. When a node receives a message, it saves the Hop-by-Hop-Identifier for correlating messages, and generates its own unique identifier before routing the message forward. Each node that receives a message follows this process, so that a unique identifier is added to the message at every hop.

When a message is received, the Hop-by-Hop-Identifier is compared with any pending requests in the queue. If there is a match, the message is removed from the queue, and the answer message is processed.

When formulating an answer to a message, the same Hop-by-Hop-Identifier that was used in the request is inserted into the answer by the node sending the answer. Every node will have saved the unique identifier when it was received, and will replace that identifier when forwarding the answer back to the destination.

End-to-End Identifier   The End-to-End Identifier is different than the Hop-by-Hop-Identifier. It is used to identify a message from the originator to the destination, and is globally unique. We will explain its use in more detail a little further down.

Processing Diameter Messages

When answering a request, the receiving node must formulate an answer based on several data from the request. The same Hop-by-Hop-Identifier that was sent from the adjacent node must be used when sending back the answer, however this is only sent as far as the adjacent node. That node will use the identifier it received by an adjacent node when the request was sent to it, and on down the path until the message reaches the origination.

The Host-ID is changed to the value of the node sending the answer back to the originator. No Destination-Host or Destination-Realm AVPs are used in the answer. The "P" bit is set to the same value as the request. The Session-ID and the End-to-End-Identifier are included with the same value as the request.

User-Name AVP   The User-Name AVP contains the Network Access Identifier (NAI) of the user requesting the service. The length can be very long in some cases, at least 63 octets. RFC 2486 recommends at least 253 octets. The format is defined in RFC 2486, which includes the user realm.

Encryption of the User-Name AVP will prevent snooping of the message by fraudsters. Using IPsec or TLS/DTLS is highly recommended, especially on the network boundaries.

Disconnecting Peer Connections

When disconnecting from a peer, the disconnecting node sends Disconnect-Peer-Request (DPR) command. The Disconnection-Reason AVP is included in the command to identify why the connection is being disconnected (much like cause codes are used in SS7). If this AVP is not sent, the peer will send periodic messages in an attempt to reestablish the connection. If a peer is not going to be sending any Diameter messages for a while, it might want to send the DPR command until it is ready to begin sending traffic again. No attempt is made to reconnect unless there is traffic to be sent to the other peer.

Transport Failure Detection

Transport failures can also result in disconnects and need to be identified as soon as possible to prevent unnecessary traffic in the network. This can be done through the use of Device-Watchdog-Requests (DWR) commands sent on a periodic basis. This command is used to continuously test connections to prevent failures being discovered when messages are being sent.

When a peer has received no traffic during a specified time period (configured in timer Tw) then DWR is sent to the suspect peer. The default value for this timer is 30 seconds, and should never be less than 6 seconds. The T_w timer is reset anytime a message is received.

Failover Procedures   In the event a peer is no longer reachable, failover procedures should be implemented immediately. For example, if the Device-Watchdog-Answer (DWA) command is not received within the specified time (provisioned by the DWR/DWA timer), failover procedures should commence.

The primary connection to a peer should not be closed until T_w expires at least twice. A DWR is not sent when the timer expires the second time, but failover is initiated on the second expiration. The connection is closed at that time and the failover process begins. When a connection fails, the peer should attempt periodically to reestablish the connection.

Failback Procedures   When failover is implemented, any messages in the queue are sent to alternative peers. This does not include DWR, which would not have any significance if sent to a secondary peer. If duplicate messages are received during failover, the End-to-End-Identifier is used along with the Origin-Host to detect duplicates.

Error Handling

When an agent receives a message with the Result-Code AVP indicating an error, the AVP cannot be altered. Only the error code recorded in the Result-Code AVP is sent to the destination.

If on the other hand an error occurs locally but the Result-Code AVP indicates success, the Result-Code AVP is altered to reflect an error at the agent and returned to the destination. The Error-Reporting-Host AVP is also added to the answer message.

Identifying Errors

In addition to connection errors, there are also protocol errors and application errors. Protocol errors happen at the base protocol level somewhere during transport. This is usually identified hop-by-hop and indicated somewhere between the originator and the destination.

Application errors occur at the application level and may not require any actions from agents in the routing path. An answer message is generated with the "E" bit set, and the Result-Code with the appropriate value sent to the originator of the request.

Error Message Format   When an error is detected, an answer message is used to communicate the error back to the originator of the request. The answer message takes on the following format:

<answer-message>::=<Diameter HDR: Code, Err>

0*1 <Session-ID>

{Origin-Host}

{Origin-Realm}

{Result-Code}

[Origin-State-ID]

[Error-Message]

[Error-Reporting-Host]

[Failed-AVP]

[Experimental-Result]

*[Proxy-Info]

*[AVP]

The Result-Code AVP is then used to identify the type of error that occurred.

Application Errors

There are multiple values for application errors, depending on the type of error that occurred. A request with an AVP that is not recognized, and listed as mandatory ("M" bit is set) will result in the Result-Code value DIAMETER_AVP_UNSUPPORTED. The Failed-AVP AVP is also sent containing the AVP that failed.

A request that contains an AVP with an unrecognized value will cause the Result-Code value to be set to DIAMETER_INVALID_AVP_VALUE. Again, the Failed-AVP AVP is sent with the failed AVP.

Any request that is missing a required AVP will generate a Result-Code with the value DIAMETER_MISSING_AVP. The Failed-AVP AVP will be sent with the AVP that was expected. If there are multiple errors, only the first error detected is reported.

Result-Code AVP   Any node reporting an error must use the Result-Code AVP to identify the error. The node reporting the error must also include its identity in the Origin-Host AVP. The Result-Code values are based on five different error classes.

1xxx—Informational

2xxx—Success

3xxx—Protocol Error

4xxx—Transient Error

5xxx—Permanent Failures

Informational Messages   Informational error codes are used to indicate a request could not be processed, and the requestor requires additional action. For example, DIAMETER_MULTI_ROUND_AUTH 1001 indicates a request for authentication requires multiple round trips, and a subsequent request needs to be sent.

Success Messages   Success messages are used to indicate a request was processed successfully. There are two values defined in RFC 6733 for this AVP; DIAMETER_SUCCESS 2001, and DIAMETER_LIMITED_SUCCESS 2002. The latter value is used to inform the requestor that the message was processed, but there is additional processing required to complete the request.

Protocol Errors   Protocol errors are reported in the Result-Code AVP sent in an answer message. The answer message is sent with the "E" (error) bit set to one and the Result-Code contains the proper value for the error. As with all error messages, the code is the same as the request code. The "P" bit in the header is set to the same value as the received request, but the "R" bit is set to a value of "0."

A protocol error is reported as soon as it is detected, at each hop where it is detected. If detected by a proxy, the proxy may be able to fix the error and forward the message to its destination, but if not, it will also return an answer with the error bit set.

The following values are supported in the AVP to identify protocol errors:

DIAMETER_COMMAND_UNSUPPORTED 3001

DIAMETER_UNABLE_TO_DELIVER 3002

DIAMETER_REALM_NOT_SERVED 3003

DIAMETER_TOO_BUSY 3004

DIAMETER_LOOP_DETECTED 3005

DIAMETER_REDIRECT_INDICATION 3006

DIAMETER_APPLICATION_UNSUPPORTED 3007

DIAMETER_INVALID_HDR_BITS 3008

DIAMETER_INVALID_AVP_BITS 3009

DIAMETER_UNKNOWN_PEER 3010

DIAMETER_TOO_BUSY indicates the receiver is unable to process the request because of congestion, and the message should be sent to an alternate peer.

When DIAMETER_REDIRECT_INDICATION is sent, it indicates the redirect agent was unable to fulfill the request and therefore the request should be redirected to the host identified in the Redirect-Host AVP.

Transient Failures   Transient failures are requests that could not be processed when received, but could be processed later. For example, if the correct password were used.

The Result-Code AVP identifies the type of transient failure that occurred. For example, DIAMETER_OUT_OF_SPACE indicates an accounting message that was received but the server is out of storage capacity and unable to accept the accounting message.

ELECTION_LOST indicates the sending peer has lost the election process during the connection phase and is disconnecting the transport.

The following AVP values are support for transient errors:

> DIAMETER_AUTHENTICATION_REJECTED 4001
>
> DIAMETER_OUT_OF_SPACE 4002
>
> ELECTION_LOST 4003

The DIAMETER_AUTHENTICATION_REJECTED value indicates the user failed authentication, typically because the user entered an invalid password. The user should then be prompted to the proper password prior to subsequent attempts.

Permanent Failures   Permanent failures indicate just that—the failure is permanent and no further attempts should be made. When a permanent failure is sent, the 'E' bit is not set in the header, except for one exception. There could be cases where messages for specific applications with applicable grammar could not be created. In these cases an answer message with the 'E' bit set would be sent with applicable grammar.

The following AVP values are supported for permanent errors:

> DIAMETER_AVP_UNSUPPORTED 5001

When this is sent, it indicates that a mandatory AVP was included in the request but the AVP is not supported. The 'M' bit is set in the request, indicating the mandatory AVP. The unsupported AVP is sent in the answer message as part of the Failed-AVP contents.

> DIAMETER_UNKNOWN_SESSION_ID 5002
>
> DIAMETER_AUTHORIZATION_REJECTED 5003

The above error most likely indicates that the user is not authorized to use the service.

> DIAMETER_INVALID_AVP_VALUE 5004
>
> DIAMETER_MISSING_AVP 5005

Indicates an AVP is missing. The Failed-AVP AVP is sent containing the missing AVP. If the AVP is proprietary and vendor specific, the Vendor-ID AVP is also sent in the answer message.

> DIAMETER_RESOURCES_EXCEEDED 5006
>
> DIAMETER_CONTRADICTING_AVPS 5007

The home server has received a message containing AVPs that contradict one another. The Failed-AVP AVP is sent in the answer message containing the contradicting AVPs.

> DIAMETER_AVP_NOT_ALLOWED 5008
>
> DIAMETER_AVP_OCCURS_TOO_MANY_TIMES 5009
>
> DIAMETER_NO_COMMON_APPLICATION 5010
>
> DIAMETER_UNSUPPORTED_VERSION 5011

As of this book, version 1 is the only version out there, so there should not be any instances where this error occurs (unless a message is referencing another version besides version 1). When the next version of Diameter is defined, it will result in a new RFC from the IETF identifying the new version.

DIAMETER_UNABLE_TO_COMPLY 5012

This is somewhat of a generic message. It allows the sender to reject a message for no specific reason.

DIAMETER_INVALID_BIT_IN_HEADER 5013

A common reason is the reserved bit is set to one, but it could be any other invalid bit in the header.

DIAMETER_INVALID_AVP_LENGTH 5014

DIAMETER_INVALID_MESSAGE_LENGTH 5015

DIAMETER_INVALID_AVP_BIT_COMBO 5016

DIAMETER_NO_COMMON_SECURITY 5017

This is sent after receiving a CER and there are no common security mechanisms. The value is sent in the Result-Code AVP contained in the CEA.

Most all of the above messages are sent along with the Failed-AVP AVP containing the AVP that generated the error. In some cases, multiple AVPs can be contained in the Failed-AVP AVP.

Error-Message AVP   This AVP can be used along with the Result-Code AVP. The text error values are designed to be read by the end users, much in the same fashion as HTTP errors are displayed to end users. If the error message is to be processed automatically rather than read by a human, than the Error-Message AVP is not needed.

Error-Reporting-Host AVP   The Error-Reporting-Host AVP is used to identify the address of the host sending the error message. This is only used for protocol, transient, and permanent errors. If the Host-ID is of the same value, then the Error-Reporting-Host AVP is not used, as the host is identified already.

Failed-AVP AVP   The Failed-AVP AVP contains the AVP(s) that failed. There can be multiple instances of failed AVPs, but typically just the first AVP to fail is returned. This limits error response to one error response at a time. If the failed AVP is a grouped AVP, then the entire group can be included, but it is not mandatory. This allows you to see where the error occurred within the grouped AVP.

Experimental-Result AVP   This AVP indicates a vendor specific AVP has failed, or was successful, depending on the type of error message being sent. The Vendor-ID must also be sent to identify the vendor for the proprietary AVP, as well as the Experimental-Result-Code AVP. This is a grouped AVP, following the format below:

<Experimental-Result> :: = <AVP Header: 297>

{Vendor-ID}

{Experimental-Result-Code}

Experimental-Result-Code AVP   The Experimental-Result-Code AVP contains a vendor defined error code, but the codes should follow the same format as those defined by the IETF. This means they should also follow the classifications. There is no risk of duplicate error codes, because the Vendor-ID and the Experimental-Result-Code AVPs must be used together, and as long as there are no duplicate codes from the same vendor, the combination ensures uniqueness.

Looping Detection   Looping is detected when a server sees its own address in the Route-Record AVP. When this happens, the Result-Code AVP is sent with the value of DIAMETER_LOOP_DETECTED. The relay or proxy agents check for looping.

Agents also check to verify the destination does not appear in the Route-Record AVP prior to sending a message. This also helps ensure looping does not occur. If looping is detected prior to sending a message, and there are no alternative destination peers, then an answer message is sent back to the originator of the request with the Result-Code set to a value of DIAMETER_UNABLE_TO_DELIVER.

Creating User Sessions

When a Diameter client wishes to send a request, it must first set the value for the Session-ID. Any subsequent messages related to the user session will use the same Session-ID. Request for authorization messages are not defined in the base protocol. They are defined by each of the applications themselves, as they are application specific.

Prior to sending a Diameter message, the sender must follow several steps. The first step is to set the appropriate command code for the request message in the header, as well as set the "R" bit (indicating this as a request).

A unique end-to-end and hop-by-hop identifier must be given to the message. The End-to-End identifier must be globally unique, but the Hop-by-Hop identifier only need be unique to the sender. This is so the answer can be correlated to the request. The origin host and origin realm is added to the message, and the IP address for the destination host and realm is added to the header. Once the message is sent to its destination, the message is placed in the pending queue until an answer is received.

If the message is vendor specific, the Vendor-ID AVP is added to identify the vendor. This AVP could include the Product-Name and Firmware-Revision AVPs for sending debugging information, or for identifying specific nodes that should be able to decipher and process a message.

Session-ID AVP   The client when creating a request sets this AVP value. The Session-ID is maintained throughout the life of the session, to ensure answers can be correlated to the original request. The value must be unique globally, as it is used to reference a session by many nodes. It must also be maintained as a unique value for historical significance. In other words, if messages are archived for accounting applications (for example), the Session-ID could be used for identifying specific messages in the archive. The Session-ID AVP always follows the message header. The ID is comprised of the following:

<Diameter Identity>; <high 32 bits>; <low 32 bits>; [<optional value>]

The Diameter identity is typically the IP address of the node (either IPv4 or IPv6) including the address type (identified in the first two octets). IPv4 is of type 1, and IPv6 is of type 2.

The high and low 32 bits are decimal numbers generated by the session initiator. When a node reboots, the node may set the high order bits to the time based on the Network Time Protocol (NTP). The low order bits are set to zero. The combination of the Diameter identity, the time, and the increasing number represented in the low order bits ensures a globally unique identifier.

The optional value is application specific, but it can be things like an identifier for a modem, a timestamp, or other locally generated value. This is not required as part of the session ID so unless an application specifies that this be used, most Session-IDs will be absent of the optional value.

An example of a Session-ID where there is no optional value:

Accesspoint1.example.com;1256891245;930

Note that the IP address is actually a URL in this case, and would need to be resolved.

Session-Binding AVP   This is used to inform the receiving node that all future reauthentication/authorization messages and STRs for a specific application must be sent to the same server. The values for this AVP are:

1 = RE_AUTH

2 = STR

4 = ACCOUNTING

If the value is RE_AUTH (1), then all reauthentication/authorization messages for this session must include the Destination-Host AVP. If the value is STR (2), than the STR message must not include the Destination-Host AVP. If the value is ACCOUNTING (4), then the Destination-Host AVP is not included for accounting messages. If this AVP is not present (default), then the Destination-Host AVP is required.

Maintaining a Session

There are several ways for maintaining sessions for a defined period of time. The application or the service host can specify a specific period that a service is to be granted, for example. This limits the exposure to abuse by consumers of the service.

One means of maintaining a session is by using the Session-Timeout AVP. Another means might require the user to be reauthorized. This might be the case for prepaid services, to ensure a user does not consume more service than they are allowed. When answering a request, the server can assign a specific period of time that the user will be granted access to the service. Below are the various AVPs used for defining the amount of time a session is to be granted.

Session-Timeout AVP   The Session-Timeout AVP identifies the number of seconds service is to be provided before the session is terminated. Both the Authorization-Lifetime AVP and the Session-Timeout AVP can be present but the Session-Timeout AVP must have a value equal to or less than the Authorization-Lifetime AVP.

When a session is terminated because the Session-Timeout has expired, an STR must be sent unless both the client and server agree to a stateless session. STR is not sent in the case of a stateless session. If no Session-Timeout is sent, or the value is zero, than the session duration is unlimited.

The client may wish to send this AVP as a hint to the maximum duration of time that the client is willing to be fiscally responsible for a session. It is not recommended that sessions be allowed unlimited durations as this opens up the network to fraud.

Authorization-Lifetime-AVP The Authorization-Lifetime AVP is used to define the maximum number of seconds a service is to be provided to a user before they have to be reauthenticated and/or reauthorized. Since this also represents the amount of time the service provider is willing to be fiscally responsible for a session, there should be careful consideration as to the value for this AVP.

However, if the value is too low, there will be an increase in Diameter traffic attributed to reauthentication messages, and this could cause congestion in the network and at the application. A value of zero indicates the access node requires immediate reauthentication/authorization. A value of all ones means no reauthentication/authorization is expected. The absence of this AVP also means no reauthentication/authorization is needed.

If the Session-Timeout AVP is also present in the message, the value of Session-Timeout cannot be smaller than the value of the Authorization-Lifetime AVP. This would result in the session being terminated prior to a reauthenticate/authorization even occurring.

When the Authorization-Lifetime AVP is included in the reauthentication/authorization messages, it indicates how many seconds going forward (from receipt of the message) that the service is to be provided.

The value can also be used as a hint to the server as to how much time the client is willing to be fiscally responsible for the session. In this case, the answer returned must contain the same value, or a smaller value, but cannot be of a greater value or the client will reject it.

Auth-Grace-Period AVP There is a grace period before a session is removed, identified in the Auth-Grace-Period AVP. This AVP defines how many seconds after the lifetime of a session that was granted expires, before a session is removed. Both the Auth-Grace-Period and the Authorization-Lifetime AVPs are important for billing systems, because they define how long the home network is willing to be fiscally responsible for a session.

Re-Auth-Request (RAR) A server may initiate reauthentication and/or reauthorization while a session is in progress. This would be used to validate the service is still being used for example. The RAR is sent to the access point, and if the Session-ID matches a session in progress, the access point must start the reauthentication/authorization process back toward the device. The Re-Auth-Answer (RAA) is used in response and communicates the success or nonsuccess of the request.

Re-Auth-Request-Type AVP This is used in application specific answers to tell the client what action is expected when Authorization-Lifetime AVP value expires. It must be present if Authorization-Lifetime AVP is sent in an answer message with a value of one. There are two possible values:

0 = AUTHORIZE_ONLY (default)

1 = AUTHORIZE_AUTHENTICATE

Multi-Round-Time-Out AVP   This AVP is used to define the amount of time an access node will wait for a response when sending a reauthentication/authorization request to the user. It contains the maximum number of seconds that it will wait for the response. It is used in application specific answers where the Result-Code has a value of DIAMETER_ MULTI_ROUND_AUTH.

Tracking Session State

The client and the server maintain session state. The stateful agent tracks all active sessions. A session is considered active until it is terminated or it expires. Expiration is determined by the value set in the Session-Timeout AVP. Some applications may define their own session state machine; however, those applications must still make use of the Session-ID, STR/STA, and ASR/ASA messages as defined by the base protocol.

The base protocol defines four state machines, two for state being maintained from the client perspective and two when state is not maintained from the servers' perspective.

If an access device does not intend to maintain session state, it will use the Auth-Session-State AVP to indicate whether or not state will be maintained. There are two possible values:

0 = STATE_MAINTAINED

2 = NO_STATE_MAINTAINED

The server will then not maintain state since no session termination is expected from the access device. The STR will not be sent as this is only used for stateful servers.

Stateless agents may also maintain the transaction state for every session. If there is a failure, the agent invokes failover procedures. A stateless agent uses the Proxy-Info AVP for inserting local state into a request. The answer should have the same value.

Each node saves the received hop-by-hop identifier and replaces it with its own unique value before forwarding the message.

When an answer is received, the receiving node enters the original value for the hop-by-hop identifier, and the message is forwarded on to its destination. The hop-by-hop identifier must be correlated therefore with the Session-ID.

For example, a node receives a Diameter request with the hop-by-hop identifier of "1" and the Session-ID of "A." Before forwarding the request, the node will save the hop-by-hop identifier for this session and replace it with Session-ID of "9." When the node receives the Diameter answer, with the Session-ID of "A" and the hop-by-hop identifier of "9," it will replace the identifier with the value of "1" and forward the answer back to the originator of the request.

Agents maintain a copy of messages awaiting an answer. This allows them to verify state when an answer is received for a request.

Authentication/Authorization State Machines   Tables 3.10 to 3.12 show the various state machines used for authentication and authorization sessions.

Accounting State Machines   Accounting applications use their own set of state machines. A default state machine is defined by the base protocol to be used by all applications that do not have a state machine defined (see state machine #2). Other applications may define their own state machines, but they also must support the base protocol functionality as a minimum.

State	Event	Action	New State
Idle	Device requests access	Send service-specific authorization request	Pending
	ASR received for unknown reason	Send ASA with Result-code value of UNKNOWN_SESSION_ID	Idle
	RAR received for unknown reason	Send RAA with Result-Code value of UNKNOWN_SESSION_ID	Idle
Pending	Successful service-specific authorization answer received with default Auth-Session-State value	Grant Access	Open
	Successful service-specific authorization answer received, but service not provided	Sent STR	Disconnect
	Error processing successful service-specific authorization answer	Sent STR	Disconnect
	Failed service-specific authorization answer received	Clean up	Idle
Open	User or client device requests access to service	Send service-specific authorization request	Open
	Successful service-specific authorization answer received	Provide service	Open
	Failed service-specific authorization answer received	Disconnect user and device	Idle
	RAR received and client will perform subsequent reauthorization	Send RAA with Result-Code value of SUCCESS	Open
	RAR received and client will not perform subsequent reauthorization	Send RAA with Result-Code value of SUCCESS. Disconnect the user and device	Idle
	Session-Timeout expires on the access device	Send STR	Disconnect
	ASR received, client will comply with the request to end the session	Send ASA with the Result-Code value of SUCCESS, then send STR	Disconnect
	ASR received. The client will not comply with the request to end the session	Send ASA with the Result-Code value of SUCCESS	Open
	Authorization-Lifetime and Auth-Grace-Period expires on the access device	Send STR	Disconnect
Disconnected	ASR received	Send ASA	Disconnect
	STA received	Disconnect the user and device	Idle

TABLE 3.10 Server Perspective, Stateless

State	Event	Action	New State
Idle	Service-specific authorization request is received, and user is authorized	Send successful service-specific answer	Open
	Service-specific authorization request is received and the user is not authorized	Send a failed service-specific answer	Idle
Open	Service-specific request is received and the user is authorized	Send successful service-specific answer	Open
	Service-specific authorization request is received and the user is not authorized to use the service	Send a failed service-specific answer and cleanup the session	Idle
	Home server wants to confirm authentication and/or authentication of the user	Send RAR	Pending
	Home server wants to terminate the service	Send ASR	Disconnect
	Authorization-Lifetime and Auth-Grace-Period expires on the home server	Clean up	Idle
	Session-Timeout expires on the home server	Clean up	Idle
Pending	Received RAA with a failed Result-Code	Clean up	Idle
	Received an RAA with Result-Code value of SUCCESS	Update session	Open
Disconnect	Failure to send ASR	Wait, then resend ASR again	Disconnect
	ASR successfully send and ASA received with Result-Code	Clean up	Idle
Not-Disconnected State	ASA is received	No action	No state change
Any State	STR is received	Send STA, clean up session	Idle

TABLE 3.11   Server Perspective, Stateful

State machine #3 can be used by accounting servers, which may encounter problems associated with long-duration connections. This would only apply to applications where the Accounting-Realtime-Required AVP value is equal to DELIVER_AND_GRANT. The user is only disconnected if there are connectivity issues. The supervisory timer T_s is used to maintain the connection. The value of T_s should be twice the value given in the Acct-Interim-Interval AVP to prevent the state from shifting to idle when short-term transient errors occur.

State	Event	Action	New State
Idle	Client or device requests access to a service	Send service-specific authorization request	Pending
Pending	Successful service-specific authorization answer is received with the Auth-Session-State value of NO_STATE_MAINTAINED	Grant access	Open
	Failed service-specific authorization answer received	Clean up session	Idle
Open	Session-Timeout expires on the access device	Disconnect the user and device	Idle
	Service to the user is terminated	Disconnect the user and the device	Idle

TABLE 3.12   Client Perspective, Stateless

"Failure to send" in the state machines has the same meaning here as in other state machines. The peer is down, there is a transient failure, or a protocol error occurred. The following protocol errors are considered as temporary failures:

- DIAMETER_OUT_OF_SPACE
- DIAMETER_TOO_BUSY
- DIAMETER_LOOP_DETECTED

Failed answer means a failure occurred, and notice was received (nontransient errors).

Accounting records can be received at any time and in any order. The processing of accounting messages is application specific so we will not get into processing of those messages here. However, Tables 3.13 to 3.15 will provide the various state machine changes as well as actions taken by both the client and the server.

The states are defined as follows:

- PendingS = Start
- PendingI = Interim
- PendingL = Stop
- PendingE = Event
- PendingB = Buffered record

Following are the state machines that apply to accounting sessions.

Origin-State-ID AVP   The Origin-State-ID AVP can be included in any message. It is usually sent when a node reboots to verify the state of sessions. The receiving node checks Session-IDs stored to see if it has any sessions with a lower value than what is provided by the originator of this AVP. If there are sessions of a lower value, the node can assume that the sessions have been lost.

The value is numerical, and increases each time the node has a reboot. The time of the reboot is typically used as the value.

State	Event	Action	New State
Idle	The client or the device request service access	Send accounting Start request	PendingS
	Client or the device request a one-time service	Send accounting event request	PendingE
	Records in storage	Send record	PendingB
PendingS	Successful accounting start received	No action	Open
	Failure to send and buffer space available and real-time not equal to DELIVER_AND GRANT	Store start record	Open
	Failure to send and no buffer space available and real-time equal to GRANT_AND_LOSE	No action	Open
	Failure to send and no buffer space available and real-time not equal to GRANT_AND_LOSE	Disconnect the user and device	Idle
	Failed accounting start answer received and real-time equal to GRANT_AND_LOSE	No action	Open
	Failed accounting start answer received and real-time not equal to GRANT_AND_LOSE	Disconnect the user and device	Idle
	User service terminated	Store, stop, and record	PendingS
Open	Interim interval lapses	Send accounting interim record	PendingI
	User service terminated	Send accounting stop request	PendingL
PendingI	Successful accounting interim answer received	No action	Open
	Failure to send, buffer space available (or old record can be overwritten) and real-time not equal to DELIVER_AND_GRANT	Store interim record	Open
	Failure to send, no buffer space available, and real-time equal to GRANT_AND_LOSE	No action	Open
	Failure to send, no buffer space available, and real-time not equal to GRANT_AND_LOSE	Disconnect user and device	Idle
	Failed accounting interim answer received and real-time equal to GRANT_AND_LOSE	No action	Open
	Failed accounting interim answer received and real-time not equal to GRANT_AND_LOSE	Disconnect user and device	Idle
	User service is terminated	Store, stop, and record	PendingI
PendingE	Successful accounting event answer received	No action	Idle
	Failure to send and buffer space available	Store the event record	Idle
	Failure to send and no buffer space available	No action	Idle
	Failed accounting event answer received	No action	Idle

TABLE 3.13   #1—Client Accounting State Machine, Stateless

State	Event	Action	New State
PendingB	Successful accounting answer received	Delete record	Idle
	Failure to send	No action	Idle
	Failed accounting answer received	Delete record	Idle
PendingL	Successful accounting stop answer received	No action	Idle
	Failure to send and buffer space available	Store, stop and record	Idle
	Failure to send and no buffer space available	No action	Idle
	Failed accounting stop answer received	No action	Idle

TABLE 3.13 #1—Client Accounting State Machine, Stateless (*Continued*)

State	Event	Action	New State
Idle	Accounting start request received and successfully processed	Send accounting start answer	Idle
	Accounting event request received and successfully processed	Send accounting event answer	Idle
	Interim record received and processed successfully	Send accounting interim answer	Idle
	Accounting stop request received and processed successfully	Send accounting stop answer	Idle
	Accounting request received, no space left to store records	Send accounting answer with Result-Code value of OUT_OF_SPACE	Idle

TABLE 3.14 #2—Server, Stateless Accounting

State	Event	Action	New State
Idle	Accounting start request received and successfully processed	Send accounting start answer. Start time T_s	Open
	Accounting event request received and processed successfully	Send accounting event answer	Idle
	Accounting request received, no space available to store records	Send accounting answer with Result-Code value of OUT_OF_SPACE	Idle
Open	Interim record received and successfully processed	Send accounting interim answer. Restart timer T_s	Open
	Accounting stop request received and successfully processed	Send accounting stop answer. Stop timer T_s	Idle
	Accounting request received, but no space left to store records	Send accounting answer with Result-Code value of OUT_OF_SPACE. Stop timer T_s	Idle
	Supervision timer T_s expired	Stop timer T_s	Idle

TABLE 3.15 #3—Server, Stateful Accounting

Class AVP   The Class AVP is used to return state information to an access node. When there are multiple Class AVPs in a message, they must be used in any subsequent reauthentication/authorization session termination and accounting messages.

The current state overrides the previously received values, if sent in a reauthentication/authorization message. The size of this AVP cannot exceed 4096 bytes. The client will terminate the session if it receives an AVP exceeding this length, or exceeds available storage.

Session Termination

When a server is maintaining state, it must receive notification when the session is being terminated for any reason. This only applies to stateful servers. The following procedures will not apply to stateless servers.

For authorization services, the access node providing the services issues a Session-Termination-Request (STR) command to the server that authorized the service. This also applies to cases where a session may not have actually been started. Transport failures or protocol errors may have prevented the session from being established. In these cases, STR is still sent to the server so it does not believe the session is still in progress.

In some cases, STR may not be received by the server, but expiration of timers will end the session. For example, the Session-Timeout, Authorization-Lifetime, or Auth-Grace-Period timers expire.

Origin-State-ID can be used for detection of terminated sessions when an STR is not sent. This could be used for example when an access node crashes. The counter is incremented every time the access node is started up, and the new value sent in CER/CEA. This value is then compared to any stored value the server may have for the node. If the stored value is lower, the server knows that the node has been restarted.

Upon detection of a client reboot, the server will clean up the status of any sessions and may send an STR to any other servers for global cleanup of sessions.

Session-Termination-Request (STR)   The STR is sent to notify a Diameter server that the session identified in the Session-ID is to be terminated, and all resources associated with the service are to be released.

The STA is sent in response to an STR to indicate that all resources associated with the Session-ID have been released. It could also indicate a failure with the presence of the Result-Code AVP and a value other than SUCCESS (error codes). The format for these commands are as follows (Table 3.16):

Termination-Cause AVP   This AVP is sent to identify why a session is being terminated. The values are defined by the IANA and RFC 6733:

1 = DIAMETER_LOGOUT

2 = DIAMETER_SERVICE_NOT_PROVIDED

3 = DIAMETER_BAD_ANSWER

4 = DIAMETER_ADMINISTRATIVE

5 = DIAMETER_LINK_BROKEN

6 = DIAMETER_AUTH_EXPIRED

7 = DIAMETER_USER_MOVED

8 = DIAMETER_SESSION_TIMEOUT

STR	STA
<Session-ID>	<Session-ID>
{Origin-Host}	{Result-Code}
{Origin-Realm}	{Origin-Host}
{Destination-Realm}	{Origin-Realm}
{Auth-Application-ID}	[User-Name]
{Termination-Cause}	*[Class]
[User-Name]	[Error-Message]
[Destination-Host]	[Error-Reporting-Host]
*[Class]	[Failed-AVP]
[Origin-State-ID]	[Origin-State-ID]
*[Proxy-Info]	*[Redirect-Host]
*[Route-Record]	[Redirect-Host-Usage]
*[AVP]	[Redirect-Max-Cache-Time]
	*[Proxy-Info]
	*[AVP]

TABLE 3.16   STR/STA Command Content

Session-Server-Failover AVP   This AVP is sent in the answer message when an STR or a reauthentication/authorization message fails because of a problem in the delivery. The answer message is application specific. The Session-Binding AVP can also be present, but it must have a value of zero.

When this AVP is received, the receiving node will try the request again without the Destination-Host AVP. The following values are supported:

0 = REFUSE_SERVICE

1 = TRY_AGAIN

2 = ALLOW_SERVICE

3 = TRY_AGAIN_ALLOW_SERVICE

If the value is REFUSE_SERVICE (0), then the service is terminated and there is no subsequent attempt to establish the service. However, if the value is TRY_AGAIN (1), the originator of the request will try sending the request again without the Destination-Host AVP in the request.

If the value is ALLOW_SERVICE (2), then the originator can assume the reauthorization/authentication succeeded even if reauthentication/authorization failed. If an STR is delivery failed, the service is terminated.

If the value is TRY_AGAIN_ALLOW_SERVICE (3), the originator of the request will attempt the request again without the Destination-Host AVP in the request. If reauthentication/authorization fails, the originator assumes reauthorization was successful. If an STR fails, the session is terminated.

Session Abort

When a session is aborted, the Abort-Session-Request (ASR) is sent. This could be used for example when the quota limit of a subscriber has been reached, and the session needs to be aborted. Stopping a session upon receipt of an ASR is implementation dependent. Some access nodes may only accept ASRs from specific agents, for example. This is highly recommended for security reasons.

The ASA is sent in response to an ASR. It will contain the Result-Code AVP indicating the action that was taken. If the session was terminated, the value is DIAMETER_ SUCCESS. If the session is not active, the Result-Code value is set to DIAMETER_ UNKNOWN_SESSION_ID. If the access node fails to stop the session for any other reason, the Result-Code value is sent as DIAMETER_UNABLE_TO_COMPLY. The receiving node must send an ASA in response, identifying the action it took using the Result-Code AVP (Table 3.17).

Security in Diameter

As networks move toward an all-IP infrastructure, security becomes more important. We have moved from networks that are owned and operated by a handful of known corporations, to a diverse ecosystem with many partners and many different types of networks. There is no longer a trust factor between network operators, and therefore, security must be implemented to maintain control of network access.

Today's protocols are open and accessible by most anyone, and all that is needed is a means of connecting. This is not actually as easy as some researchers have claimed. However, today many exploits between roaming partners have been discovered, further increasing the need for security at the network boundaries.

This is best achieved by establishing a network security domain. One administrator, such as a network operator, maintains a security domain. In some cases, operators may

ASR	ASA
<Session-ID>	<Session-ID>
{Origin-Host}	{Result-Code}
{Origin-Realm}	{Origin-Host}
{Destination-Realm}	{Origin-Realm}
{Destination-Host}	[User-Name]
{Auth-Application-ID}	[Origin-State-ID]
[User-Name]	[Error-Message]
[Origin-State-ID]	[Error-Reporting-Host]
*[Proxy-Info]	[Failed-AVP]
*[Route-Record]	*[Redirect-Host]
*[AVP]	[Redirect-Host-Usage]
	[Redirect-Max-Cache-Time]
	*[Proxy-Info]
	*[AVP]

TABLE 3.17   ASR/ASA Command Content

want to establish subdomains. For example, an operator with networks in multiple countries should operate each network as a separate security domain.

Security domains are interconnected through security gateways (SEGs). Security gateways are distributed throughout the network boundaries to protect against attacks and unauthorized access. It is usually a good idea to implement a security gateway with limited access to network resources, to limit attack surfaces. The security gateway is only used to protect the control plane and not the user plane.

The 3GPP has defined IPsec as the security protocol to be used at the SEG. IPsec provides a number of services, but the minimal services to be provided include

- Data integrity
- Authentication of data origin
- Antireplay protection
- Confidentiality

When using IPsec, the security gateways will establish a security association. The association is defined by security policy. While IPsec can offer a number of different services, and different security protocols, 3GPP has defined encapsulated security payload (ESP) to be used in 3GPP networks.

Key management is supported through the use of Internet Key Exchange (IKEv1) and Internet Key Exchange (IKEv2). The role of IKEv1 and IKEv2 is to negotiate, establish and maintain security associations between two domains.

A security association is identified by three parameters:

1. Security parameter index (SPI)
2. IP destination address (the ESP security association address)
3. Security protocol identifier (always ESP)

Security policy is what defines the profile for all traffic passing between networks. The security policy is maintained in a security profile database, and is queried each time traffic is to be passed through the SEG. This is what tells the SEG how the traffic is to be handled from a security perspective. Some traffic may be allowed to pass in the clear, while other traffic may be defined for protection.

Security policies implemented between the network element and the security gateway should be consistent throughout the security domain. In other words, all network elements that need to pass traffic to another network should be given the same security policy when connecting to the security gateway. Security policy between networks is something that must be negotiated by the interconnecting operators and is defined in the roaming agreement.

3GPP has defined IPsec per the IETF specifications, but to simplify the implementation, not all aspects of IPsec are required in 3GPP networks. Many of the IPsec options are not recognized in 3GPP networks. However, there are some options in the IPsec suite that have been deemed mandatory by 3GPP. Refer to 3GPP TS 33.210 for more details.

The GSMA has also defined the requirements for the interconnection of wireless networks using the Diameter protocol. IR-88 is the latest document outlining the steps for securing these networks, and it should be noted that adherence to IR-88 will ensure that interconnects are not abused.

The most fundamental requirement is the use of IPsec at the edge of the network. While many service providers will opt not to use encryption within their networks, they should not opt-out of using encryption on the network boundaries.

This is especially true when nodes are exchanging capabilities through the use of CER/CEA. Sending this command without encryption will expose the network provider. Encryption should be either TLS/DTLS or IPsec, both recognized by the IETF. The GSMA and the 3GPP have both defined IPsec as the security protocol of choice.

Another measure toward securing the network is the use of IP access lists (ACL). An access list is a table of IP addresses that are allowed access to the network. The ACL may be inherent in the IP router, or it may also be part of the routing translations in a Diameter agent (such as a Diameter router). This is where static IP addressing should be used, ensuring the service provider always knows who they are connecting.

Topology hiding is also a requirement of the GSMA. Topology hiding relies on obfuscation to prevent the topology of the network form being discovered. Prior to sending a request or an answer, the Diameter Edge Agent (DEA) will replace the host name and realm to another value, usually random.

This requires the DEA to remember what name and realm it used for the message, so when dealing with responses or any associated messages it can correlate them to the real name and realm. There are a number of other techniques used here, and this is not meant to be an exhaustive description of topology hiding so check with your Diameter router vendors for details on how your equipment supports this feature.

These are the minimal requirements for securing the network. Many of the network exploits being seen today can be attributed directly to the lack of these simple measures. The industry once relied on "trusted" partners to support roaming and other services, but this is no longer such a model, and no network should be trusted.

CHAPTER 4

Routing in the Diameter Network

Diameter Agents

The Diameter protocol uses agents to route messages through the network. There are three basic types of agents:

1. Relay agents

2. Redirect agents

3. Proxy agents

There is also a Diameter Edge Agent as defined by the Groupe Speciale Mobile Association (GSMA), which is used at the network boundaries to support interconnection of Diameter networks. The GSMA has defined a minimal set of functionality to be supported by edge agents, such as encryption and topology hiding.

The industry uses a lot of different names for these network elements. Some vendors will use the term "Diameter signaling controller" while others will call them "Diameter routing agents (DRAs)," although the latter is incorrect given the specific definition for a DRA. Oracle uses the Diameter Signaling Router as their product name, which is probably more accurate given the many features and functions it supports.

Regardless of the label given by vendors, the functionality breaks down into these three basic functions, and can be supported in the same network node.

Relay Agents

The relay agent receives Diameter messages, and then routes them to their destination based on the routing information in the message header. The agent does not need to read the contents of the message, other than the Diameter header for routing information. The relay agent also maintains state of each transaction, but do not maintain the state of a session.

To perform this task, the relay agent must have a list of Diameter peers that it is able to route messages to. The Application-ID AVP is used to determine the application that must be supported by the destination, so Application-ID becomes an important part of the routing decision.

They may also have the ability to modify the contents of the routing information in the header, but they cannot modify any other data in the message. For example, the relay agent must append the header with the Route-Record AVP, containing the address of the previous node.

This is different than Record-Route in SIP, where each node appends its own IP address. In Diameter, the agents add the IP address of the node they received the message from, providing another layer of routing information.

Relay agents are useful for aggregating many connections into a hub and spoke network model. This helps eliminate a lot of complexity in the network architecture, while providing a central location for monitoring and security functions.

Another important role of the relay agent is to look and detect looping in the network.

Redirect Agents

Redirect agents provide the information needed to properly route a message to its destination. If a message is routed to a relay agent, and it does not know how to route the message, it may forward the message to a redirect agent for further instructions.

When a server receives a redirected message, it adds the message to its message queue and forwards to the destination. If no connection exists for the destination, the server establishes the connection. The destination could be another redirect agent.

The redirect agent will route the message with the "E" bit in the header set to a value of 1. The Result-Code AVP is added to the message with the value of DIAME-TER_REDIRECT_INDICATION. The hop-by-hop identifier is maintained as received, rather than the agent resetting the counter. The Redirect-Host AVP is also added with the address of the redirect agent used in routing the message. Every redirect agent will add its own separate Redirect-Host AVP.

If an answer message is received containing the Redirect-Host-Usage AVP and the value is other than zero, a route entry is saved locally with the value found in the AVP. This entry is saved for a period of time as specified in the Redirect-Max-Cache-Time AVP.

There can be multiple entries for these routes, so they must be resolved based on the usage. One route may have a value of ALL_SESSION, while another may have the value of ALL_REALM. If the value is ALL_SESSION, any message received with this in the Redirect-Host-Usage AVP is routed to the designated peer. If a message contains this AVP with the value of ALL_REALM, the message would be routed to the peer designated for this routing entry.

Redirect agents do not need to read the entire contents of the message. They only need to see the message header for routing instructions.

Redirect agents do not maintain session state, but they do maintain transaction state.

Proxy Agents

Proxy agents also route based on the header information, but it also may need to read the entire contents of a message to understand the services and applications that must be supported. This is needed to determine the best network node to receive the message.

An example of this might be DRA as defined by the standards. The DRA reads the contents of a message before selecting the proper Policy and Charging Rules Function (PCRF) to route the message to.

This also allows a proxy to provide local processing in some cases, if the proxy supports the application that is needed. They advertise the applications they support through the Application-ID AVP.

The proxy agent routes messages based on the network address in the header. Looking at the realm portion of the network address indicator (NAI), as well as the state of downstream peers, the proxy agent is able to make routing decisions in the event of failed network elements.

Proxy agents also detect looping in the network, and also append messages using the Record-Route AVP.

Diameter Path Authorization

Encryption is used for Diameter connections as a means of authenticating at the transport layer. This is something that Signaling System #7 (SS7) networks have been unable to provide, and in later years may have proven a problem for SS7 networks. Using TLS or IPsec, Diameter is able to authenticate the connection; preventing replay attacks while protecting network integrity.

Prior to establishing a connection with peers, Diameter performs a capability exchange. This determines what applications the node is able to support and is maintained by other connecting nodes in their peer tables.

For example, a relay agent must advertise it supports the relay application during the capability exchange. This ensures messages to be routed forward are only sent to relay agents. They advertise their support for applications using the Capabilities Exchange Request/Answer (CER/CEA) commands, and the Application-ID AVP.

Realm-Based Routing

An agent can only route to a specific host in the network if that host is listed in the nodes peer table. If the peer is not listed in the peer table, the Destination-Realm and the Application-ID AVP are used for routing. The Destination-Realm can also be derived from the User-Name AVP.

Nodes should have a list of all supported realms and application IDs as well as externally supported realms and application IDs in a peer table. Routing should be based on these provisioned peer tables rather than peer discovery, due to security vulnerabilities when peer discovery is used.

Originating a Request

When sending a request, the originator of the request must identify itself using the Origin-Host AVP. The originator is the only one that can enter this AVP, and relay agents cannot modify the value.

Likewise the Origin-Realm is required in all messages to identify the network or realm the message was originated in. Routing decisions are made based on these two AVPs.

The Destination-Host AVP is found in all messages that are sent by an agent but it is not necessarily found in requests, and is never included in answers. Destination-Realm is also used for routing decisions when included in a message. This AVP is not found in answer messages, but can be found in messages originated by agents. It is derived from the User-Name AVP by clients.

Receiving Requests

When a message is received, and the Destination-Host contains the address of the receiving node, the node processes the message (since it is addressed to that node). If there is no Destination-Host address in the message, but the Destination-Realm is the same as that of the receiving node, the node will assume the message is being routed to it and will process the message. The exception is if the receiving node does not support the application requested.

Routing AVPs

There are some AVPs that are used specifically by the Diameter agent for the purpose of routing a message through the Diameter network. These AVPs contain information about reaching the destination (as would be the case of a proxy for example).

Route-Record AVP

A node adds the Route-Record AVP when it receives a message. It contains the address of the node that the message was received from, derived from the Origin-Host contained in the CER/CEA.

The value of this AVP is used for sending responses. They should follow the same path as the request. This is a means of preventing man-in-the-middle attacks, and is similar to the approach used in the SIP protocol.

This is another way of ensuring that messages are only routed through trusted networks. A service provider may elect not to send answer messages through certain networks if the request took the path of untrusted networks. If the service provider does not wish to grant access to a service because the request was received through unknown or untrusted networks, the receiver can return an answer message with DIAMETER_MESSAGE_REJECTED in the Result-Code AVP.

Record-Route is also used for loop detection. See Chap. 3 for more information on how loop detection works.

Proxy-Info AVP

The Proxy-Info AVP contains the identity and the local state of the node that creates a message. This is a grouped AVP and contains the Proxy-Host AVP (host identity) and the Proxy-State AVP (local state of the host).

<Proxy-Info> :: = <AVP Header: 284>

{Proxy-Host}

{Proxy-State}

*{AVP}

Proxy-Host AVP

The Proxy-Host is the host adding the Proxy-Info AVP to a message. This will be in the format of an IP address, or name/realm of the host.

Proxy-State AVP

The host that provides this information will also provide the state of the sending host. Other nodes that receive this AVP, unless they are maintaining the state of the proxy, may not use this information. Refer to the description of state machines in Chap. 3 for more information about state machines and how this information is used.

Peer Tables

Peer tables are used to determine how to route to a Diameter peer. When a node receives a message, and the Destination-Host is listed in the peer table, the message is forwarded to the peer. The peer table must identify the following for each peer:

- Host-Identity
- Status-T (must be one of the following values)
  - Closed
  - Wait-I-CEA
  - Wait-Conn-Ack/Elect
  - Wait-Returns
  - R-Open
  - I-Open
  - Closing
- Static or dynamic
- Expiration time
- TLS/DTLS enabled

The Host-Identity is derived from the contents of the Origin-Host AVP sent in the CER/CEA. The static or dynamic status indicates how the peer was entered into the table. If the value is static, it was manually entered. If the value is dynamic, it was entered via peer discovery.

The expiration time specifies how long the entry is to remain before it is refreshed or deleted from the peer table. This is only used with dynamic entries, for maintaining the entries.

Routing tables differ in that the peer table will point to a routing table for network routing of a message. The routing table contains the following information:

- Realm name
- Application ID
- Local action (describes the action to be taken through one of the following values)
  - Local
  - Relay
  - Proxy
  - Redirect
- Server identifier
- Static or dynamic
- Expiration time

The realm name is the first "key" used in routing. Think of the realm as the same thing as a domain name. The realm identifies the network or sub-network that the message is to be routed to. The Application-ID is the second key used in routing.

The Application-ID must be used to find the right server supporting the application being requested. For example, the Home Subscriber Server in the Long Term Evolution network is an application, and will be used for authentication as well as authorization requests.

The receiver uses the value of local action to determine if there is anything that should be done by the receiving node. For example, a value of "local" indicates the receiving node should process the message.

Static or dynamic is the same as in the peer table, and used to indicate if the route is the result of a manual entry or one determined by peer discovery. The author does not recommend dynamic routes, as these can be a security risk. The expiration time applies only to dynamic routes.

CHAPTER 5

Diameter for Accounting

T he Diameter protocol was designed from the beginning to provide real-time delivery of charging information for both online charging and offline charging. SCTP ensures that delivery is successful with minimal packet loss, and no fragmentation.

Batched mode accounting is not supported by Diameter; however, if there is heavy traffic, records can be grouped together in a single message rather than several smaller messages. This reduces the traffic load.

Accounting Architecture

The base protocol (RFC 3688) only defines two interfaces for accounting, but there are now several interfaces defined by the 3GPP. Here we will be identifying three of the main interfaces used to connect to the billing systems.

There is offline charging (think of this as postpaid billing) and there is online (or prepaid). Much of the world uses a prepaid billing model, where charges are recorded throughout a session, and quota is assigned for each session. Quota is considered the amount of credit that a subscriber has available, and can be increased by alerting the subscriber that they have exhausted their quota and providing them a means for "topping off" their account.

When this is supported, Diameter messages must be used over the accounting interfaces to capture the transaction, communicate the new quota (or credit) available to the various access nodes in the network supporting sessions, and terminating the session when quota has been depleted (or when the session is over).

The policy and charging rules function (PCRF) is responsible for communicating between the access nodes and the billing systems. The 3GPP defined a new interface for this given the amount of transactions between the PCRF and the online charging system (OCS). This is the Sy interface, and allows the PCRF to connect with the billing system and manage quota for various sessions based on real-time transactions while the user sessions continue.

Each network element that is responsible for reporting accounting records has a charging trigger function (CTF). This applies to both offline and online charging architectures. The CTF sends accounting records to the charging data function (CDF) in the OCS, or to the online charging function (OCF).

The CTF can only send accounting records to an accounting server within the same realm or domain. This means that network elements cannot send accounting records to accounting servers in other networks, across network boundaries. This would represent a serious security vulnerability given the sensitive information being sent in accounting records.

Routing of Accounting Messages

Accounting records can be routed two ways: split or coupled. Split accounting uses a centralized accounting server for all applications. The Application-ID AVP is the same regardless of the application being supported, indicating the application is accounting.

Coupled accounting uses the same server providing the application. The application server may then forward accounting records to an accounting server. In this case, the Application-ID does not indicate the accounting application, but contains the value of the actual application providing the service.

For example, if SMS were being billed through split accounting, the accounting server would receive the billing records. The Application-ID would indicate the application is accounting.

However, if coupled accounting were used, the same Short Message Service Center (SMSc) would receive accounting records. The Application-ID would indicate SMS as the application, and the SMSc would send the messages to the accounting server. The default value is always split accounting.

Sending Accounting Records

Even though the clients are originating accounting messages, the server can control how accounting records are sent to the accounting servers. The server controls the clients' behavior through the use of two AVPs.

The Acct-Interim-Interval AVP provides instructions for continuous accounting records even while a session is in progress. This AVP is sent in the answer to an accounting request. The Accounting-Realtime-Required AVP provides instructions for sending accounting records when delays of transport failures occur. The accounting server may change instructions during a session by sending an answer message with these AVPs. The client than uses the new values for all future accounting messages.

Clients use the base Diameter protocol failover procedures as defined in RFC 6733, and protect against record loss and duplicate records. The Session-ID and Accounting-Record-Number AVP are used to check for duplicate accounting messages.

Clients save accounting records in a buffer until an answer message is received. Buffering of messages helps to prevent message loss. The messages are held in a buffer until an answer is received. In the event that buffers become full, the oldest accounting messages are deleted to make room for the newest messages.

Correlating Accounting Records

Looking at the answer message that established the session should validate accounting requests. If there is no answer message, the accounting message should be scrutinized. There should be some mechanism for validating sessions and the accounting for those sessions, especially in cases where the service delivered was different than the service requested. In the event a service cannot be delivered, the cause code AVP is sent with the value of DIAMETER_UNABLE_TO_COMPLY.

Sometimes, billing for multiple applications is managed by one billing record. In other words, a subscriber could be using multiple applications such as text messaging and a data session, but they are both billed as one accounting record.

This requires the ability to correlate the accounting records from the different applications. The Acct-Multi-Session-ID AVP provides an identifier for both the applications to use. The value for this identifier will be the same for both of the applications in the above example. The AVP is generated with the Accounting-Request (ACR) command, and the value sent must be used in all subsequent accounting records for the session. The actual value is application specific, but it must remain the same throughout the session. The value must also be globally unique, as it will be used historically for correlating accounting records. This suggests maybe the time and date could be part of the identifier value.

Rf Interface for Offline Accounting

The Rf interface is used for both session-based charging and event-based charging in an offline model. It is recommended that a charging proxy be used to provide routing to available accounting servers, and to ensure that accounting records are not lost during transport failures. A charging proxy has the ability to save charging records should a connection be lost or the accounting server cannot be reached for any reason.

Rf Procedures

There are two different types of accounting: event based and duration based. Event-based accounting occurs whenever there is a billable event, such as a text message is sent. Duration-based billing is time based.

When using event-based accounting, an ACR is sent with the Accounting-Record-Type AVP set to EVENT_RECORD. The ACR is sent when the service has been delivered to the CDF, where the accounting record is stored. Since time is not involved there is no reason for subsequent records for the session. The accounting records simply record an event took place, such as a text message being sent.

If time-based accounting is used, then multiple ACRs are used. The accounting starts when the Accounting-Type-Record AVP is set to START_RECORD. When the session is over, the ACR is sent with the Accounting-Type-Record AVP set to STOP_RECORD. The STOP_RECORD is only sent once for a session, and results in the termination of the session in accounting.

When sending accounting records during a session, the Accounting-Type-Record AVP is sent with the value of INTERIM_RECORD. The Acct-Interim-Interval AVP provides instructions as to what intervals the interim billing records will be sent. START_RECORD and INTERIM_RECORD can be sent in the same message, as can INTERIM_RECORD and STOP_RECORD, but they should have the same Accounting-Sub-Session-ID. The application decides how the accounting records will be sent.

There is also an identifier for tracking the interim records. The Accounting-Sub-Session-ID AVP is used to track the interim records and correlate them at the accounting server. The Session-ID is still used to identify the session, but the START_RECORD, INTERIM_RECORD, and STOP_RECORD messages carry the same Accounting-Sub-Session-ID. An Accounting-Sub-Session-ID can have only one START_RECORD and STOP_RECORD.

The CDF is responsible for the generation of Call Detail Records (CDRs) based on the charging information received on the Rf interface from the CTF. The CDF maintains a timer for session-based charging that is reset every time an ACR with INTERIM_RECORD or STOP_RECORD is received.

If a connection to the CDF fails, the accounting record is sent to the secondary or backup CDF. The charging proxy can also buffer messages until a connection is restored. No reply from the CDF results in the ACR being retransmitted continuously based on configurable timers that define the transmission intervals and the maximum number of retransmission attempts.

Rf Commands

There are two commands used on the Rf interface. We have already defined the CER/CEA command and its use in the base protocol. CER/CEA is used in the accounting network to exchange capabilities during the connection phase, as defined in the base protocol.

Accounting-Request/Answer

The Accounting-Request (ACR) command is used to send accounting information after a session has been authorized. The command is acknowledged by sending the Accounting-Answer (ACA) command. As with other commands and applications, if an error occurs, the answer message will contain the error code in the Result-Code AVP.

The ACR command must include the Session-ID AVP (as is true of all commands) and the User-Name AVP. User-Name is only required if it is available. A typical time-based session will generate several ACR/ACAs to record the entire session time, and the associated charges. The amount of ACRs/ACAs and the intervals they are sent is determined by the accounting server, and communicated to the CTF in the ACA.

Connecting to Online Charging Systems

Online charging as mentioned above ensures services are paid in advance, before the services are actually used. A credit is given to the subscriber account, and that credit (or quota) will be "allocated" for the session. Usually the policy and charging rules function (PCRF) is responsible for reporting the quota with the OCF, and communicating to the "enforcement points" when quota has been exhausted. This is typically performed using the Sy interface.

If the PCRF plays a role in delivery of services in online charging architectures, there is an ongoing dialog between the PCRF and the OCS during sessions, with the PCRF acting as the creator of rules to be applied to sessions based on configurable rules in the PCRF. The PCRF communicates with the OCS using Sy.

Ro Interface

The Ro interface is used for "prepaid" accounting records. Online charging uses the concept of a quota, given to a subscriber based on the balance on their account. When their quota expires, the accounting server can either require additional funds be added to the account, or service can continue until the session is terminated, and the subscriber is denied additional services until they "top-up" their account.

The Ro interface connects policy enforcement points (or access points) to the accounting server (or the OCF in this case). This interface may not be used if a PCRF is present, allowing enforcement points to connect to the PCRF instead. The PCRF than communicates with the OCF to determine the actions to be taken for a session.

Ro Procedures

Online charging supports three different scenarios; immediate event charging (IEC), event charging with reservation, and session charging with reservation. A reservation is a unit of credit or quota being allowed for the session, based on the available funds in the subscribers account. When a subscriber uses their entire quota, reauthorization is required.

The OCF supports two subfunctions: rating and unit determination. The OCF can be centralized or it can be part of the CTF in the network element itself. For larger networks, having a centralized OCF function makes more sense.

Rating converts units to a monetary value. The units are determined by the unit determination. They can be values such as data volume, time, or events. They are determined prior to service delivery.

The OCF determines the quota limits for a session based on the type of service being requested and the subscribers' profile. The subscribers profile is usually stored in a database called the subscriber profile repository (SPR). This profile will include the type of data plan that a subscriber is entitled to, as well as services they may or may not be allowed to access.

If the CTF is going to be responsible for unit determination, it must determine the number of units it will make available and request those units from the OCF. The OCF then checks the account balance for the subscriber before granting those units to the CTF. The CTF is responsible for managing those units and the service delivery.

The rating function converts those units into a monetary value therefore the rating function need to be in the OCF. The CTF can determine the units it wants to deliver, but the value of those units is determined by the rating function in the OCF.

For example, a device requests a service such as a data connection. The access point (SGSN) providing the data connection determines the unit to be granted (megabytes). The SGSN CTF sends the Charging-Control-Request (CCR) to the OCF requesting a number of units for the session.

The OCF in turn must determine the value of those units (rating function) and then check the subscriber account to ensure they have adequate credit for the requested units. If there is enough credit on account, the OCF will send the Charging-Control-Answer (CCA) granting the requested units. However, if the account does not have enough credit, the CCA will be sent denying the service because of insufficient funds. The CTF must then request another value of units or the PCRF may request the subscriber be notified to "top-up" their account and direct them to an account portal.

Immediate Event Charging   Immediate event charging (IEC) is used to bill for events, such as a text message being sent, or maybe a ring tone being downloaded. The procedures are a little different than with the other methods of charging.

When a device requests a service, and it is determined that the service will use immediate event charging, the CTF sends to the OCF the units to be debited. The rating function is assumed in this case to be centralized, while the unit determination can be either centralized or located with the service.

The OCF checks to make sure there is enough credit in the subscribers account based on the rating (calculating the monetary value of the service units). If there is enough credit on account, the OCF will send the CCA to the CTF to notify it that the account has been debited and the service can be delivered. Note that the account is debited immediately, prior to service delivery.

The CTF can also perform the rating function, depending on the configuration of the network. Rating and unit determination can be distributed across the access nodes, or centralized.

Event Charging with Unit Reservation Reservation works a little differently. When a device requests service, the CTF will request units to be reserved, but the account is not charged until service delivery is complete. The CTF requests the reservation by sending the amount of units to be reserved to the OCF in the CCR command.

The OCF rates the units, validates the account has sufficient credit, and reserves the units from the account. The CTF is then notified of the reservation through the CCA command. It is the responsibility of the CTF to monitor the consumption of the granted units by the device. The CTF can deliver the service all at once, or in fractions, depending on the type of service offering.

When the credits have been consumed, the CTF sends another CCR to the OCF to have the units deducted from the account. If additional units are required, the CTF can reserve those units, but if no additional units are required, then the account is debited and the session is closed.

Again, the exact flow may differ slightly depending on where unit determination and rating is done. The CTF can perform unit determination and rating, or those functions can be centralized. Unit determination can also be done in the CTF while rating is performed in the OCF.

Session Charging with Unit Reservation The only difference between session charging and event charging in these examples is the service being charged. In session charging, the subscriber is being billed based on the type of session and usually duration of the session (or amount of bandwidth consumed). Event charging is based on a one-time event (such as a download).

The process is the same regardless of using event charging or session charging. CTF can perform unit determination, rating, or both, and likewise, the OCF can perform these functions as well. The service design determines how the service will be billed to the subscriber.

Some sessions may have to be reauthorized when quota has been consumed. The server may inform the CTF when to reauthorize a session for additional quota. This prevents too much quota being granted at one time, opening up the opportunity for fraud if the subscriber has large amounts of credit on account.

Direct debiting with IEC Direct debiting is similar to unit reservation, except no reservation is made. The account is debited as soon as the service is requested. The CTF requests the number of units to be granted, and the rating function determines the monetary value to be debited from the account. The CCR will have the CC-Request-Type AVP value of EVENT_REQUEST, while the Requested-Action AVP is set to DIRECT_DEBITING.

The OCF will return the CCA to the CTF, with Granted-Service-Unit and Cost-Information AVPs. There may also be the Remaining-Balance AVP present in the CCA.

Note that since the service is delivered after the account is debited, there is a chance that the subscriber does not use all of the units granted. When this happens, a refund must be credited to the subscribers account.

For example, if the service was granted, but the service delivery failed, a credit will be due back into the subscribers account. This is achieved by sending the Requested-Action AVP in a CCR to the OCF with the value of REFUND_ACCOUNT. The CTF will also include the Refund-Information AVP in the CCA.

The OCF then credits the account based on the received CCR, and returns a CCA to the CTF to let it know that the account has been credited. The CTF sets timer Tx when it sends the CCR, and resets the timer once the CCA is received from the OCF.

Ro Commands and AVPs

The Ro interface uses the CER/CEA during the connection phase, but once the connection is established, accounting records are exchanged using the Charging-Control-Request/Answer (CCR/CCA). The main AVPs are described in this section.

Credit-Control-Request/Answer (CCR/CCA)   As we saw in the procedures section, it is the CCR/CCA that is used to request credit for a specific service. The CCR and CCA are used in quota management, and are different than the ACR/ACA used in offline charging (Table 5.1).

Accounting AVPs

There are a number of AVPs used on both the Ro and Rf interfaces. Those AVPs are all listed in this section.

Accounting-Input-Octets AVP (Code 363)   Used together with the Accounting-Input-Packets AVP, this parameter tells how many octets were sent during a specified charging interval. It specifies the volume of data uploaded during a session and used for accounting on the Ro or Rf interfaces (either OCS or OFCS).

Accounting-Input-Packets AVP (Code 365)   This AVP is used in conjunction with the Accounting-Input-Octets AVP. This AVP reflects the number of packets sent during the charging interval. The size of the packets is reflected in the Accounting-Input-Octets AVP. This is used on the Ro and Rf interfaces for accounting.

Accounting-Output-Octets AVP (Code 364)   Used with the Accounting-Output-Packets AVP to identify the number of octets sent on a downlink during the charging interval. The value in this AVP identifies the size of the packets identified in the Accounting-Output-Packets AVP, so these two AVPs need to be paired. They are used on the Ro and Rf interfaces.

Accounting-Output-Packets AVP (Code 366)   This is used with the Accounting-Output-Octets AVP, which identifies the size of the packets being sent. This AVP identifies how many packets were sent in the downlink direction during the charging interval.

Accounting-Realtime-Required AVP   This is sent by the accounting server or authorization server to provide instructions to the client on what to do if it cannot send accounting records (due to a network problem for example). The values are

1 = DELIVER_AND_GRANT

2 = GRANT_AND_STORE

3 = GRANT_AND_LOSE

CCR	CCA
<Session-ID>	<Session-ID>
{Origin-Host}	{Result-Code}
{Origin-Realm}	{Origin-Host}
{Destination-Realm}	{Origin-Realm}
{Auth-Application-ID}	{Auth-Application-ID}
{Service-Context-ID}	{CC-Request-Type}
{CC-Request-Type}	{CC-Request-Number}
{CC-Request-Number}	[CC-Session-Failover]
[Destination-Host]	*[Multiple-Services-Credit-Control]
[User-Name]	[Cost-Information]
[Origin-State-ID]	[Low-Balance-Indication]
[Event-Timestamp]	[Remaining-Balance]
*[Subscription-ID]	[Credit-Control-Failure-Handling]
[Termination-Cause]	[Direct-Debiting-Failure-Handling]
[Requested-Action]	*[Redirect-Host]
[AoC-Request-Type]	[Redirect-Host-Usage]
[Multiple-Services-Indicator]	[Redirect-Max-Cache-Time]
*[Multiple-Services-Credit-Control]	*[Proxy-Info]
[CC-Correlation-ID]	*[Route-Record]
[User-Equipment-Info]	*[Failed-AVP]
*[Proxy-Info]	[Service-Information]
*[Route-Record]	*[AVP]
[Service-Information]	
*[AVP]	

TABLE 5.1   CCR/CCA Command Content

If the value is one (1), DELIVER_AND_GRANT, the service is granted only if there is a connection to the accounting server (including backup servers), which are treated as one server. If the value is two (2), GRANT_AND_STORE, the service is granted if there is a connection to an accounting server, or there is sufficient storage capacity to prevent loss of records.

If the value is three (3), GRANT_AND_LOSE, the service will be granted even if accounting records cannot be sent or stored. Since no accounting records will be processed for the service, this value represents a loss of revenue.

Accounting-Record-Number AVP   This AVP is used to identify a single accounting record so it can be correlated with other associated accounting records. For example, when interim accounting records are being sent for a session, the Session-ID and the Accounting-Record-Number AVP together will create a globally unique identifier for the accounting record during correlation.

Accounting-Record-Type AVP   This AVP is used to identify the type of accounting record that is being sent. Remember there are two types of accounting records; an event record and a time-based record. The values for this AVP are

1 = EVENT_RECORD

2 = START_RECORD

3 = INTERIM_RECORD

4 = STOP_RECORD

Accounting-Sub-Session-ID AVP   This AVP correlates interim records for the designated session, identified by the Session-ID. If this AVP is absent, then subsessions are not being used. A STOP_RECORD without this AVP will terminate all subsessions.

Acct-Application-ID AVP (Code 259)   Used to identify numerically the application this message is supporting. When used on the Ro and Rf interface, the application identifier is accounting, which has a numeric value of 3.

Acct-Interim-Interval AVP   This defines how often accounting records are to be sent. If this AVP is absent, or if it contains a value of zero (0), the START_RECORD, INTERIM_RECORD, and STOP_RECORD are generated based on the service being provided.

The INTERIM_RECORD is sent at regular intervals if the value is greater than zero (0). The value will define the interval in seconds for sending interim accounting records. When other sessions are producing accounting records, there could be a situation where many accounting records are being sent during the same intervals. The value is randomized to prevent this from happening. A proxy can also be used to prevent congestion from large volumes of accounting records.

Acct-Multi-Session-ID AVP   This AVP is used to correlate accounting messages that are related, but have different Session-IDs. It would typically be returned in an authorization answer message, and must be in every accounting message for the associated session.

Acct-Session-ID AVP   This is only used when translating between RADIUS and Diameter. It contains the value of the Acct-Session-ID parameter used from the RADIUS message.

Auth-Application-ID AVP (Code 258)   When this AVP contains a value of 4, it indicates the application this message is being used for is authentication/authorization.

Called-Station-ID AVP (Code 30)   Identifies the Access Point Name (APN) that the user is connected to for the specified session.

Event-Timestamp AVP (Code 55)   Marks the time that a chargeable event was received by the CTF. Used on the Ro and Rf interface. This is sent in the Accounting-Request and Accounting-Answer messages to provide the time an event occurred. The time is recorded in seconds, with the timer starting on January 1, 1900, 00:00 UTC.

Multiple-Services-Credit-Control AVP (Code 456)   This is a grouped AVP, consisting of the following:

<Multiple-Services-Credit-Control>::=<AVP HDR: 456>

[Granted-Service-Unit]

[Requested-Service-Unit]

*[Used-Service-Unit]

[Tariff-Change-Usage] (not used in 3GPP)

*[Service-Identifier]

[Rating-Group]

*[G-S-U-Pool-Reference]

[Validity-Time]

[Result-Code]

[Final-Unit-Indication]

[Time-Quota-Threshold]

[Volume-Quota-Threshold]

[Unit-Quota-Threshold]

[Quota-Holding-Time]

[Quota-Consumption-Time]

*[Reporting-Reason]

[Trigger]

[PS-Furnish-Charging-Information]

[Refund-Information]

*[AF-Correlation-Information]

*[Envelope]

[Envelope-Reporting]

[Time-Quota-Mechanism]

*[Service-Specific-Info]

[QoS-Information]

*[AVP] (not used in 3GPP)

The various AVPs in this grouping provide more detailed information about the charges to be applied for multiple services that were delivered for the session. This is used on the Ro and Rf interfaces.

Rating-Group AVP (Code 432)   When a quota is granted for a session, a rating group "key" is assigned to the quota. This is a unique value used to identify the assigned quota during charging. This AVP contains that charging key.

Result-Code AVP (Code 268)   The result code AVP is used to identify failures or success in a transmission. There are transient failures, and permanent failures that have been defined by 3GPP for use on Ro and Rf interfaces, in addition to the result codes defined in the base protocol RFC 6733. The additional values for transient failures are

DIAMETER_END_USER_SERVICE_DENIED 4010

DIAMETER_CREDIT_CONTROL_NOT_APPLICABLE 4011

DIAMETER_CREDIT_LIMIT_REACHED 4012

Result code 4011 indicates that services can be delivered, but online charging does not apply. Can also be used to indicate offline charging is to be applied. Additional values for permanent failures are

DIAMETER_AUTHORIZATION_REJECTED 5003

DIAMETER_USER_UNKNOWN 5030

DIAMETER_RATING_FAILED 5031

Result code 5031 indicates an insufficient rating input, an incorrect AVP combination was received, or there was an unrecognized AVP value. The Failed-AVP AVP must also be used containing the AVP that caused the failure.

Service-Context AVP This AVP identifies the 3GPP document that defines the service being supported or identified for a session. The format for this is as follows:

Extensions.MNC.MCC.release.service-context@domain

The first parameter, "extensions," is service provider specific and not defined. The "MNC.MCC" parameter is the mobile network code and mobile country code. Release corresponds to the release of the specified document (such as release 1.0).
The values for "service-context@domain" are

32251@3gpp.org Packet switch charging

32252@3gpp.org Wide area network LAN (WLAN) charging

32260@3gpp.org IMS charging

32270@3gpp.org MMS service charging

32271@3gpp.org LCS service charging

32272@3gpp.org POC service charging

32273@3gpp.org MBMS service charging

32274@3gpp.org SMS service charging

32275@3gpp.org MMTel service charging

32280@3gpp.org Advice of Charge (AoC) service charging

Service-Identifier AVP (Code 439) This AVP is used with the Service-Context-ID AVP to uniquely identify a specific service related to a request.

Used-Service-Unit AVP (Code 446) Contains the number of measured units from the start of the service delivery. This is a grouped AVP consisting of the following AVPs:

<Used-Service-Unit>::= <AVP HDR: 446>

[Reporting-Reason]

[Tariff-Change-Usage]

[CC-Time]

[CC-Money] (not used by 3GPP)

[CC-Total-Octets]

[CC-Input-Octets]

[CC-Output-Octets]

[CC-Service-Specific-Units]

*[Event-Charging-Timestamp]

*[AVP] (not used by 3GPP)

User-Name AVP (Code 1)   Contains the name of the network address indicator (NAI) of the user as described in RFC.

Validity-Time AVP   This AVP is used to communicate the value of timer Tcc and is controlled by the OCS.

Sy Interface (PCRF to OCS)

Since the PCRF is responsible for quota management in most networks, the 3GPP thought it prudent to define a direct interface between the PCRF and the OCS. In earlier standards, the PCRF had to communicate through the enforcement points (such as an SGSN or PDN Gateway) which really does not make much sense.

The OCS maintains a set of "policy counters" which can be used to represent any form of unit for any type of service. These counters are assigned to each subscriber starting a session. The PCRF will then subscribe to these counters so it can receive status changes from the OCS while the subscriber session is connected.

For example, a subscriber may have a specific amount of money on their account. When they initiate a data session, the SGSN will ask the PCRF what rules need to be applied to the session. The PCRF will then connect with the OCS, check the balance on the account (a policy counter) and return this information to the SGSN (over simplified to keep it simple at this point).

The PCRF communicates with the OCS throughout a session, so it will establish an Sy session with the OCS for each subscriber session. The OCS acts as a Diameter server, while the PCRF acts as the client.

Establishing an Sy Connection

Each session is for a specific user and user session. Note that a user can be a subscriber, or another network, or any type of device that is accessing the network for services and will be billed for that access. Session state is always maintained on Sy, so the Auth-Session-State AVP is not used on this interface when a session is established.

The PCRF is responsible for linking its Gx or S9 session with the Sy session. The Gx or S9 sessions are with access nodes managing the user connection to services.

These are also known as policy enforcement points in the 3GPP standards for policy and charging. These policy enforcement points are where the decisions made by PCRF are actually enforced, since this is where the "bearer path" actually connects. The PCRF therefore can only create the rules, use the OCS updates to understand the status of the users account, and send those rules via the Gx or S9 interface to its enforcement points. The PCRF will maintain the association of the Gx/S9 interface with the associated Sy interface sessions until the user access is released, or the user connection no longer relies on OCS interaction.

Sy uses either TCP or SCTP for transport of Diameter messages to the OCS. The accounting functions defined in the beginning of this chapter do not apply on the Sy. Accounting functions are supported on the interfaces between the access nodes and the OFCS/OCS directly. Sy is used for maintaining online accounting with quota management.

To establish a connection with PCRF or OCS on the Sy, either node must first advertise its capabilities during the capabilities exchange. The originator of the session request sends the Supported-Features AVP, listing the features supported by the host.

Since Sy is defined by the 3GPP, the 3GPP is considered the vendor. This is communicated in the Vendor-Specific-Application-ID grouped AVP. The value for the 3GPP is always 10415. The Sy application is identified in the Auth-Application-ID AVP as 16777302.

It is assumed that all nodes will support 3GPP Rel. 11 as defined in the standards. The Supported-Features AVP is really only needed if there are features not defined in Rel. 11 (considered as extensions to the standard). If the connecting nodes support Rel. 11, there is no need for the Supported-Features AVP.

Sy Procedures

The PCRF makes its policy decisions based on the status of policy counters maintained in the OCS. These counters can represent a number of different types of units, such as amount of money left on account, number of text messages allowed, or time granted for a free service. The PCRF must receive the updates to these values from the OCS for use in its decision process.

A policy counter can represent spending for one service or multiple services, one device or multiple devices, one subscriber or a group of subscribers. The standard does not define what the policy counter represents, as this is configured by the service provider in their OCS.

The PCRF subscribes to a policy counter (or a group of counters) to receive regular status updates. The PCRF uses the Spending-Limit-Request/Answer command set to subscribe. The OCS will use the Spending-Status-Notification-Request/Answer command set to provide status updates once the PCRF is subscribed.

Initial/Intermediate Spending Limit Reporting

There are three steps in the process. Initial/intermediate, status limit reporting, and final status limit reporting. The OCS reports on the status of each of the counters. The PCRF uses the status for its decision making in how rules are to be applied to specific services. The actual enforcement is maintained at the access nodes themselves.

When a subscriber starts a session, the access node will send a request to the PCRF asking for the rules that should be applied to the requested connection. The PCRF will in turn send a request to the OCS using the Spending-Limit-Request command over the

Sy interface, for policy counters to be assigned for the subscriber session. The SL-Request-Type AVP is sent in the request with a value of INITIAL_REQUEST.

The OCS will return an answer once it has created the session. If this is an initial request, but there is already a session with the Session-ID, the OCS will return an error of DIAMETER_INVALID_AVP_VALUE. The Failed-AVP AVP is also sent containing the SL-Request-Type AVP with the value INITIAL_REQUEST.

If there is no session with the Session-ID, and the SL-Request-Type does not equal INITIAL_REQUEST, the OCS will send an answer with the Result-Code value of DIAMETER_UNKNOWN_SESSION_ID.

If the PCRF sends a request with unknown policy counter values, the OCS will return an error using the Experimental-Result-Code AVP with a value of DIAMETER_ERROR_UNKNOWN_POLICY_COUNTERS. Again the Failed-AVP AVP is sent in the answer message with the unknown policy counters.

If this is an initial request, no Sy session is created. If this is an intermediate request, any changes received are ignored and the original counter status is maintained.

If the initial or intermediate request does not identify any specific policy counters, the OCS will send the status of all policy counters and possibly their activation times as well.

Spending Limit Reporting

Once the PCRF has subscribed to reporting from the OCS, the OCS will use the Spending-Status-Notification-Request command to send status updates to the PCRF. When the status of a counter changes, the OCS sends this command to the PCRF with the new value of the counter, and possibly the activation time for the new value if applicable. If several counters change status at the same time, the OCS can send all of the counter changes in one message, using multiple Policy-Counter-Status-Report AVPs, one for each counter.

The PCRF will send the Spending-Status-Notification-Answer command to the OCS upon successful receipt of the request with the Result-Code of DIAMETER_SUCCESS. The PCRF then uses the received status for making decisions about rules to be applied to the service.

Final Spending Limit Reporting

When a subscriber is done with a service and terminates their connection, the PCRF will unsubscribe to the policy counters assigned to the user session. The Session-Termination-Request command is sent by the PCRF to the OCS with the Termination-Cause AVP value of DIAMETER_LOGOUT.

The OCS will then remove all of the subscriptions specified in the request. Note though that the counters do not get erased. OCS maintains account status for each subscriber, so it must maintain the status for all user accounts in the network. Only the Sy session is effected by this command.

Sy Commands and AVPs

When a user establishes a connection to a service in the network, the PCRF is used to determine what rules will be applied to the service. The PCRF works directly with the OCS to determine how much of a service the user will be allowed. There are many examples that come to my mind.

If a subscriber is roaming and they are granted free access to the network for 1 day, something in the network must track that subscribers network access and how much time is left. This is not something that can be done on the access node itself, since the subscriber may be roaming and using various access nodes within the network.

The OCS is centralized, and is the best place for granting and tracking this time. However, the access nodes need to know when the time has expired. This is the purpose of the PCRF. The PCRF will communicate with the OCS when the service is granted, and will provide updates to the access nodes themselves no matter where the subscriber is connected. Of course, this is just one example of service allocation. The value tracked in the OCS can be monetary, time, or any other units defined for a specific service.

There are two commands that are used on the Sy interface for supporting PCRF. The PCRF uses the Spending-Limit-Request command to request a subscription to a specific policy counter (or group of counters). The OCS will then use the Spending-Limit-Answer command to provide a response.

The OCS will then use the Spending-Status-Notification-Request command to send updates to the PCRF for the counters it has subscribed to, related to a specific user and user session. The PCRF uses the Spending-Status-Notification-Answer command to acknowledge receipt of the update.

The Sy interface also uses a number of the AVPs found in the base protocol, but there are some AVPs that are unique to just the Sy interface. Those AVPs and their use are defined below.

Spending-Limit-Request/Answer

The Spending-Limit-Request/Answer (SLR/SLA) command set is used by the PCRF to request a subscription to specified policy counters in the OCS. Anytime there is a change to these counters, the OCS will report the status change to the PCRF (Table 5.2).

SLR	SLA
<Session-ID>	<Session-ID>
{Auth-Application-ID}	{Auth-Application-ID}
{Origin-Host}	{Origin-Host}
{Origin-Realm}	{Origin-Realm}
{Destination-Realm}	[Result-Code]
[Destination-Host]	[Experimental-Result]
[Origin-State-ID]	*[Policy-Counter-Status-Report]
[SL-Request-Type]	[Error-Message]
*[Subscription-ID]	[Error-Reporting-Host]
*[Policy-Counter-Identifier]	*[Failed-AVP]
[Logical-Access-ID]	[Origin-State-ID]
[Physical-Access-ID]	*[Redirect-Host]
*[Proxy-Info]	[Redirect-Host-Usage]
*[Route-Record]	[Redirect-Max-Cache-Time]
*[AVP]	*[Proxy-Info]
	*[AVP]

TABLE 5.2 SLR/SLA Command Content

Spending-Status-Notification-Request/Answer

The Spending-Status-Notification-Request/Answer (SNR/SNA) command set is used by the OCS to communicate to the PCRF changes in status for specific policy counters that the PCRF has subscribed. Each policy counter is specific to a user session, and can be any form of unit such as monetary value or time (Table 5.3).

Sy Specific AVPs

There are a few AVPs that are specific to the Sy interface and not used anywhere else. These AVPs are specific to policy and policy counters.

Pending-Policy-Counter-Change-Time AVP This AVP provides the time when a pending counter status becomes the current status using Network Time Protocol (NTP). This AVP can be part of the Policy-Counter-Status-Report AVP.

Pending-Policy-Counter-Information AVP This is a grouped AVP providing the time and status of a policy counter. For instance, for a specific service, the time allotted for accessing that service might reach a threshold at a specific time, which is communicated to the PCRF in this AVP. When that time is reached, the value automatically becomes the new value of the counter.

<Pending-Policy-Counter-Information> ::= <AVP HDR: 2905>

{Policy-Counter-Status}

{Pending-Policy-Counter-Change-Time}

*{AVP}

This AVP can be part of the Policy-Counter-Status-Report AVP.

SNR	SNA
<Session-ID>	<Session-ID>
{Origin-Host}	{Origin-Host}
{Origin-Realm}	{Origin-Realm}
{Destination-Host}	[Result-Code]
{Auth-Application-ID}	[Experimental-Result]
[Origin-State-ID]	[Origin-State-ID]
*[Policy-Counter-Status-Report]	[Error-Message]
*[Proxy-Info]	[Error-Reporting-Host]
*[Route-Record]	*[Redirect-Host]
*[AVP]	[Redirect-Host-Usage]
	[Redirect-Max-Cache-Time]
	*[Failed-AVP]
	*[Proxy-Info]
	*[AVP]

TABLE **5.3** SNR/SNA Command Content

Policy-Counter-Identifier AVP   The Policy-Counter-Identifier is how the PCRF identifies the counters that it wants to subscribe to in the OCS. The policy counters are assigned to specific services, and each time a user accesses those services, a policy counter gets assigned. It is this counter that the OCS will report on, and can be a timer, a monetary value, or any other unit as defined for the specific service.

This AVP is part of the Policy-Counter-Status-Report AVP.

Policy-Counter-Status AVP   This AVP provides the value of the counter, so it will be a number (depending on the type of unit). The service provider defines the format (time, monetary value, etc.) when the service is created (during the service design). This is the value that will be reported to the PCRF.

This AVP is part of the Policy-Counter-Status-Report AVP.

Policy-Counter-Status-Report AVP   This is a grouped AVP, providing all of the parameters to the PCRF for a specific counter. It will include the AVPs defined above.

<Policy-Counter-Status-Report> ::= <AVP HDR: 2903>

{Policy-Counter-Identifier}

{Policy-Counter-Status}

*[Pending-Policy-Counter-Information]

*[AVP]

SL-Request-Type AVP   This AVP identifies what type of status request is being sent; initial, intermediate, or final.

0 = Initial request

1 = Second or subsequent request

Reused AVPs on Sy
The Experimental-Result AVP is used on the Sy to provide some additional result codes not defined in the base protocol. For permanent failures on the Sy, the value of DIAMETER_USER_UNKNOWN is used when a session is being requested for an unknown subscriber. If the OCS does not recognize the policy counters specified in a request, it will send the value DIAMETER_ERROR_UNKNOWN_POLICY_COUNTERS.

A transient failure might occur if the OCS does not have any available policy counters for a specific subscriber. The OCS returns the code DIAMETER_ERROR_NO_AVAILABLE_POLICY_COUNTERS.

All other errors use the result code values defined in the RFC 6733.

CHAPTER **6**

Connecting to Subscriber Databases

here are a number of databases used throughout wireless networks. Subscriber profiles, authentication and authorization credentials, and subscriber location all require a database to store the information needed by the network to function properly. In this chapter, we will look at the most common applications defined for connecting to these network databases.

We are only focusing on to select few databases most commonly found in LTE networks. These are the Home Subscriber Server (HSS) and the Equipment Identity Register (EIR). When connecting to these databases via Diameter, there are defined procedures used. We will examine those procedures in each section first, and then break out the protocol and the format of the commands used in each application.

Connecting to the HSS

The HSS is the most commonly used database in LTE networks. This is where information about the subscriber connecting to the network is stored. The HSS contains data about the type of subscription the subscriber has, the type of devices allowed, and the type of access allowed. It is also the HSS that stores information about the location of the subscriber when they are roaming. This information is then used in a variety of ways by the network.

Likewise, other networks need to connect to the subscribers' home network and the HSS assigned to the subscriber to be able to support roaming for the subscriber. This roaming interconnect is what allows other networks to know the services and features that a subscriber is allowed to use, and how they should be treated within the visited network.

Indeed the HSS is probably the most active interconnection and the one that requires the most protection. There are a number of network elements that connect with the HSS, but the most common are the Mobility Management Entity (MME) and the serving GPRS support node (SGSN) (Fig. 6.1).

S6a/S6d Application ID 16777251

The S6a/S6d applications are the most common applications used in wireless LTE networks. The S6a is used by the MME to exchange subscriber data with the HSS, while the S6d is used by the SGSN. The MME and the SGSN are usually combined into one

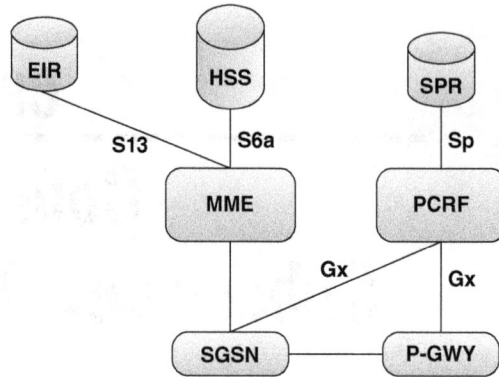

FIGURE 6.1 The evolved packet core (EPC) showing only Diameter interfaces.

network element but operate as two separate logical functions. Throughout this chapter we make the assumption they are two separate functions, although there will be some instances where we will discuss treatment of combined nodes.

S6a/S6d Procedures

Since the MME or the SGSN do not maintain the state of their sessions with the HSS, session termination is not needed. This means that the Session-Timeout and Authorization-Lifetime AVPs are not used for this application. The transport used is SCTP.

You should consider the HSS as the "keys to the kingdom." The database should be closely protected and measures used to prevent unauthorized access to the HSS. This network function holds a lot of important information that is used to either allow or not allow access to your network and access to your network services. This includes cryptographic keys used during the authentication process.

This is done through a number of different procedures used by this application. The MME or the SGSN need to know the host and realm address for the HSS. However, when connecting to outside networks, it is not desirable to provide this level of addressing to external networks. An HSS proxy should be used when connecting to external networks to protect the identity of the HSS and prevent unauthorized access.

Routing to the HSS from an external network should be done via the Destination-Realm. The Diameter edge agent (DEA) then acts as a proxy and routes the commands to the HSS. When the HSS originates a message to the MME or the SGSN, it provides its Destination-Host and Destination-Realm identity in the message. This address is then stored in the MME and the SGSN for future messages generated toward the HSS. The Destination-Realm is mandatory in all requests.

The following procedures are defined for the S6a/S6d application:

- Authentication procedures
- Mobility management
- Cancel location
- Purge user equipment (UE)
- Subscriber data handling

- Delete subscriber data
- Fault recovery
- Notification procedures

Authentication Procedures   Authentication is one of the primary procedures on S6a/S6d. The MME or SGSN use this application to request authentication from the HSS, using the Authentication-Information-Request/Answer (AIR/AIA) commands.

When a device connects to the network and requests services, the network nodes managing the connection (the SGSN and the MME) will go to the HSS to first authenticate the device, and then identify the services allowed by the device. This is based on the subscription identity, usually the International Mobile Subscriber Identity (IMSI). The IMSI is used throughout these procedures to identify a subscription, but note that this does not necessarily identify the specific device. One IMSI could have multiple devices. Each device is identified by an International Mobile Equipment Identity (IMEI), configured by the manufacturer. The S6a/S6d application is to provide the authentication credentials assigned for the subscription as part of the authentication process.

The MME or the SGSN generate an AIR to receive authentication credentials for a device. There are a number of drivers for starting this procedure but we will not go into here. If the MME is a standalone MME and not combined with an SGSN, it will include the Requested-EUTRAN-Authentication-Info AVP in the AIR. It can also request this information being returned immediately, by including the Immediate-Response-Preferred AVP. This may be the case when there are a number of AIRs being sent to the HSS, resulting in delays in responses.

The SGSN would send the Requested-UTRAN-GERAN-Authentication-Info AVP, and also has the ability to request an immediate response using the Immediate-Response-Preferred AVP. If the SGSN is combined with the MME, both the Requested-UTRAN-GERAN-Authentication-Info and the Requested-EUTRAN-Authentication-Info AVPs can be included, but the two functions process the response separately.

When the HSS receives the AIR, it will process the request by instructing the Authentication Center (AUC) to generate the KASME vectors requested. The HSS sends the generated KASME to the requesting MME, but not to the SGSN. The HSS can also identify the type of nodes requesting authentication when requesting the credentials from the AUC to enable the generation of sequence numbers when processing the request.

Mobility Management   The S6a/S6d applications are also used for mobility management. The location of the device is maintained in the HSS using the location update procedure. It is also used to update other subscription data within the HSS. The HSS receives these updates from the MME and the SGSN, and sends instructions to other MMEs and SGSNs regarding the subscription.

For example, if a subscriber is roaming, the identity of the serving network and the MME in the serving network is maintained in the HSS. If the subscriber moves to another MME, that information is updated in the HSS by the now serving MME. The HSS must then notify the previously serving MME to cancel the location of the subscriber and remove its data from its memory, as that MME is no longer serving the subscriber.

Mobility procedures are used when the device initially attaches to the network (when it powers up, for example), when it moves between serving areas (tracking updates), or the HSS resets, forcing the network to reestablish radio contact with the device and register with the HSS.

The S6a/S6d application uses the Update-Location-Request/Answer (ULR/ULA) commands for mobility management. The identity of the serving MME is found in the Origin-Host AVP in the ULR received by the HSS.

When the HSS receives a location update message, it first checks the IMSI in the update message to determine if the IMSI is a known IMSI or not. If the IMSI is not known by the HSS, the HSS will return an answer containing the Result-Code AVP with a value of DIAMETER_ERROR_USER_UNKNOWN. This is found in the ULA command.

If the IMSI is a known IMSI, but the subscriber does not currently have a connection established with the packet network, the HSS returns the Result-Code AVP with the value DIAMETER_ERROR_UNKNOWN_EPS_SUBSCRIPTION.

The HSS will also check the radio access type (RAT) to verify the method used to access the network is allowed for the subscriber based on the received IMSI. If the RAT is not allowed for this subscriber, the Result-Code AVP is set to DIAMETER_ERROR_ RAT_NOT_ALLOWED.

Last but not least, the HSS checks to ensure barring has not been set for this subscriber. If the HSS finds that barring has been set for the subscription, it will return the Result-Code AVP in the ULA set to DIAMETER_ERROR_ROAMING_NOT_ALLOWED.

When the HSS receives an ULR via the S6a application (meaning it was sent by an MME), it will send a message to the MME that was previously serving the subscriber to cancel the subscription in that MME. The Cancel-Location-Request/Answer (CLR/ CLA) commands are used to remove subscriptions from the MME on the S6a and the SGSN on the S6d.

The HSS works a little differently with the S6d application. If ULR is received by the SGSN on S6d, the HSS can send the CLR or an SS7 MAP Cancel Location message to the previous location. The HSS can also delete the stored SGSN address and number if the single registration indication flag was set in the ULR.

If the update is successful, and the HSS is not returning a Result-Code AVP in the ULA, the HSS includes the subscription data in the ULA command. The exception is if the "skip subscriber data" indication was received in the ULR.

Cancel Location Procedures   The Cancel-Location-Request (CLR) is sent by the HSS to the MME or SGSN to cancel a location update. When the MME or the SGSN receives the CLR, they check the IMSI to make sure it is a known IMSI. If the IMSI is not known, the Result-Code AVP is returned set to DIAMETER_SUCCESS. There is no need to send an error message in this case, since the procedure is being used to clear the subscription for the specific IMSI from the node. If the IMSI is not known, it was already cleared. If the IMSI is known, the subscription is cleared from the node.

There are cases where the MME and the SGSN functions are combined in a single node. When this is the case, the Application-ID AVP identifies the application (S6a or S6d), and the function associated with the application processes the message. In other words, if the CLR is sent via S6a to a combined node, only the MME function in the combined node will process the CLR.

Purge UE Procedures   If a device is inactive for a period of time, or the network operator removes a subscription from the network, the purge procedure is used to notify the HSS. The HSS is notified by the MME or SGSN that the subscribers profile has been removed from the node either manually or due to inactivity. The Purge-UE-Request/ Answer (PUR/PUA) commands are used for this procedure.

The PUR is sent by the MME or the SGSN to the HSS to notify the HSS of the purge. The HSS sets the profile to "UE purged in MME" or "UE purged in SGSN" depending on the situation. There is an indicator in the commands to notify the MME or SGSN how to handle the M-TMSI (MME) and P-TMSI (SGSN). If the "freeze M-TMSI" or "freeze P-TMSI" flags are set in the AVPs, the MME or the SGSN will not reuse these identities right away. They are reserved for a period of time before they are reused again.

When the HSS receives the PUR, it will check the Origin-Host AVP to make sure the value matches the address of the MME or the SGSN presently stored as the serving entity. If they match, the HSS will instruct the MME or the SGSN to freeze the M-TMSI and P-TMSI respectively.

Subscriber Data Management Procedures   Subscriber data must be managed in several places, and the subscriber data management procedures are defined for this purpose. The Insert-Subscriber-Data-Request/Answer (IDR/IDA) commands are used for managing the subscriber data in the network nodes (MME and SGSN). Insert-Subscriber-Data-Request (IDR) is sent by the HSS when it needs to change or update subscription data in the MME or SGSN.

For example, the network is capable of initiating measurements by the radio access network so that the network design can be optimized. We will not go into detail here about how network measurements are taken, but the process is initiated using the IDR command. When the IDR is received by the MME or the SGSN, it first checks the IMSI. As long as the IMSI is a valid and known IMSI, the MME or the SGSN will augment the subscriber data for the IMSI as received in the IDR. This could also initiate the start of other procedures in the network, such as the trace procedure where measurements are pulled from the device by the access network.

If the SGSN access is restricted to a specific coverage area, the SGSN will note this in the IDA using the IDA-Flags AVP set to "SGSN Area Restricted."

Deleting Subscriber Data   Subscriber data can be deleted as well. The Delete-Subscriber-Data-Request/Answer (DSR/DSA) commands are used by the HSS to instruct the MMS and SGSN to delete subscriber data for the specific IMSI. The command can be used to delete all the subscriber data or just parts of the subscriber data from the nodes.

For example, this command can be sent to the MME to instruct the MME to remove the session transfer number for single radio voice call continuity (SRVCC) in a Voice over LTE (VoLTE) service. When the MME or the SGSN receive the DSR, they check the IMSI. If the IMSI is valid and known by the node, the corresponding data will be deleted from the node.

The SGSN must also check regional data if it is serving specific regions. It checks its routing areas and ensures that the regional data reflects an allowed region for the subscription. If the subscription for the specified IMSI is not allowed in the routing area specified in the subscription data, the SGSN returns a DSA indicating "SGSN Area Restricted."

A subscription may have bearer channels assigned when the DSR is received. If this is the case, the SGSN will release the bearer channels and the device associated with the subscription will be disconnected from any sessions. In some cases, charging characteristics may be removed from the subscription (such as when changing a subscription from postpaid to prepaid). The SGSN should always assign a default value in place of the removed value to prevent fraudulent activity on an account, or inadvertent revenue loss.

Fault Recovery Procedures   If the HSS fails or reboots for any reason, the state of the subscriptions stored and maintained in that HSS will most probably be lost. This means it is necessary for the network to force the various devices connected to the network to be authenticated again. This is done by fault recovery procedures used by the HSS to notify the MME and the SGSN. The HSS uses the Reset-Request/Answer (RSR/RSA) commands for this procedure.

The HSS notifies the MME that it may have lost the MME identity for a subscription, and likewise notifies the SGSN that it may have lost the SGSN identity. This means the HSS cannot maintain subscriber data such as location for a subscriber and is unable to send a Cancel-Location-Request (LCR), since it no longer knows the identity of the node presently serving a subscriber.

When sending this command to the MME and the SGSN, the HSS may also send a list of subscriber IDs, rather than send individual messages for each subscription. This would be the normal case if the restart affected a group of subscribers.

When the MME or the SGSN receive an RSR, they mark the subscription as "Subscriber Restored in HSS." These nodes to determine which subscriptions may have been affected use the HSS address. The process will be triggered the next time the device connects to the network.

Notification Procedures   Occasionally, there is the need to notify the HSS of changes in the subscriber's location when an update location is not initiated. This same procedure is used when there is a change in the assignment, or removal of a serving PDN Gateway. The Notify-Request/Answer (NOR/NOA) commands are used for this procedure.

This command is issued by the MME or the SGSN and sent to the HSS. The first check is always the IMSI to ensure the IMSI value is valid and known to the HSS. If the IMSI is known and valid, the HSS will store the new information.

S6a/S6d Commands and AVPs

There are a number of commands and AVPs that are defined for this application. Each command is defined in the following section.

Update-Location-Request/Answer   The Update-Location-Request (ULR) is always sent by the MME or the SGSN to the HSS to update the location information for a specific subscriber. This allows the HSS to maintain the location of a subscriber even while they are roaming. The Update-Location-Answer (ULA) is always sent from the HSS to the MME or SGSN in response to the ULR (Table 6.1).

Cancel-Location-Request/Answer   The Cancel-Location-Request/Answer (CLR/CLA) command is sent to remove a subscriber profile from an MME and/or SGSN when the HSS receives an ULR from another MME/SGSN (Table 6.2). The HSS will indicate if this is an HSS subscription withdrawal or the result of a mobile device roaming and moving to another serving node.

The HSS can require the mobile device to go through a reattachment by indicating in the CLR that the MME or the SGSN are to detach the mobile device. This in turn would force the mobile device to reattach, and in doing so be authenticated again.

ULR	ULA
<Session-ID>	<Session-ID>
[Vendor-Specific-Application-ID]	[Vendor-Specific-Application-ID]
{Auth-Session-State}	[Result-Code]
{Origin-Host}	[Experimental-Result]
{Origin-Realm}	[Error-Diagnostic]
[Destination-Host]	{Auth-Session-State}
{Destination-Realm}	{Origin-Host}
{User-Name}	{Origin-Realm}
[OC-Supported-Features]	[OC-Supported-Features]
*[Supported-Features]	[OC-OLR]
[Terminal-Information]	*[Supported-Features]
{RAT-Type}	[ULA-Flags]
{ULR-Flags}	[Subscription-Data]
[UE-SRVCC-Capability]	*[Reset-ID]
{Visited-PLMN-ID}	*[AVP]
[SGSN-Number]	*[Failed-AVP]
[Homogeneous-Support-of-IMS-Voice-Over-PS-Sessions]	*[Proxy-Info]
[GMLC-Address]	*[Route-Record]
*[Active-APN]	
[Equivalent-PLMN-List]	
[MME-Number-For-MT-SMS]	
[SMS-Register-Request]	
[SGs-MME-Identity]	
[Coupled-Node-Diameter-ID]	
*[AVP]	
*[Proxy-Info]	
*[Route-Record]	

TABLE 6.1   ULR/ULA Command Content

Authentication-Information-Request/Answer   This command is sent from the MME or the SGSN to the HSS for authenticating a mobile device. When the HSS has validated the IMSI sent by the MME or SGSN, it will then request the AuC to generate the authentication credentials to be used. These are then sent in the AIA to the requesting node (Table 6.3).

Insert-Subscriber-Data-Request/Answer   This command is used to communicate changes to the subscriber profile (Table 6.4). It can be sent by either the MME or the SGSN to the HSS, or by the HSS to the MME or SGSN. For example, if a subscriber is connected to an MME, and there was an administrative change to the subscription profile in the HSS

CLR	CLA
<Session-ID>	<Session-ID>
{Auth-Session-State}	[Vendor-Specific-Application-ID]
{Origin-Host}	*[Supported-Features]
{Origin-Realm}	[Result-Code]
{Destination-Host}	[Experimental-Result]
{Destination-Realm}	{Auth-Session-State}
{User-Name}	{Origin-Host}
*[Supported-Features]	{Origin-Realm}
{Cancellation-Type}	*[AVP]
[CLR-Flags]	*[Failed-AVP]
*[AVP]	*[Proxy-Info]
*[Proxy-Info]	*[Route-Record]
*[Route-Record]	
[Vendor-Specific-Application-ID]	

TABLE 6.2 CLR/CLA Command Content

AIR	AIA
<Session-ID>	<Session-ID>
[Vendor-Specific-Application-ID]	[Vendor-Specific-Application-ID]
{Auth-Session-State}	[Result-Code]
{Origin-Host}	[Experimental-Result]
{Origin-Realm}	[Error-Diagnostic]
[Destination-Host]	{Auth-Session-State}
{Destination-Realm}	{Origin-Host}
{User-Name}	{Origin-Realm}
[OC-Supported-Features]	[OC-Supported-Features]
*[Supported-Features]	[OC-OLR]
[Requested-EUTRAN-Authentication-Info]	*[Supported-Features]
[Requested-UTRAN-GERAN-Authentication-Info]	[Authentication-Info]
{Visited-PLMN-ID}	*[AVP]
*[AVP]	*[Failed-AVP]
*[Proxy-Info]	*[Proxy-Info]
*[Route-Record]	*[Route-Record]

TABLE 6.3 AIR/AIA Command Content

IDR	IDA
< Session-Id >	< Session-Id >
[Vendor-Specific-Application-Id]	[Vendor-Specific-Application-Id]
{Auth-Session-State}	*[Supported-Features]
{Origin-Host}	[Result-Code]
{Origin-Realm}	[Experimental-Result]
{Destination-Host}	{Auth-Session-State}
{Destination-Realm}	{Origin-Host}
{User-Name}	{Origin-Realm}
*[Supported-Features]	[IMS-Voice-Over-PS-Sessions-Supported]
{Subscription-Data}	[Last-UE-Activity-Time]
[IDR- Flags]	[IDA-Flags]
*[Reset-ID]	[Vendor-Specific-Application-Id]
*[AVP]	*[Supported-Features]
*[Proxy-Info]	[Result-Code]
*[Route-Record]	

TABLE 6.4 IDR/IDA Command Content

(such as operator determined barring being removed) the HSS would send this to the serving node. Other reasons for this command could include

- The network has activated subscriber tracing in the MME or the SGSN.
- The HSS has requested to be notified when the mobile device has become reachable.

Delete-Subscriber-Data-Request/Answer   This command is used to remove all or a portion of a subscriber profile from the MME or the SGSN (Table 6.5). Specifically, it is used under the following circumstances:

- There is a need to remove all or a portion of the EPS subscription data for the subscriber from either the MME or the SGSN.
- The regional subscription data needs to be removed.
- There are changes to the charging profile for the subscriber. For example, the subscriber is moving from a prepaid charging model to postpaid.
- The session transfer number used for supporting SRVCC has changed or needs to be removed.
- Trace data collected by the network is to be removed.

Purge-Ue-Request/Answer   If a mobile device has been idle for a period of time (days in most cases), the subscriber profile is purged from the serving nodes. The MME or the SGSN would then send this command to the HSS to notify the HSS that the subscription has been purged.

DSR	DSA
<Session-Id>	<Session-Id>
[Vendor-Specific-Application-Id]	[Vendor-Specific-Application-ID]
{Auth-Session-State}	*[Supported-Features]
{Origin-Host}	[Result-Code]
{Origin-Realm}	[Experimental-Result]
{Destination-Host}	{Auth-Session-State}
{Destination-Realm}	{Origin-Host}
{User-Name}	{Origin-Realm}
*[Supported-Features]	[DSA-Flags]
{DSR-Flags}	*[AVP]
*[Context-Identifier]	*[Failed-AVP]
[Trace-Reference]	*[Proxy-Info]
*[TS-Code]	*[Route-Record]
*[SS-Code]	
*[AVP]	
*[Proxy-Info]	
*[Route-Record]	

TABLE 6.5   DSR/DSA Command Content

The HSS will send a PUA with PUA-Flags to indicate whether or not the M-TMSI/ P-TMSI is to be frozen (blocked from reuse for a period of time) (Table 6.6).

Reset-Request/Answer   When the HSS has a failure, the subscription records may become corrupted, or be erased altogether. This means that the subscription record between the HSS and the MME are no longer in sync. The MME must update the subscription records when requests are received for service.

The HSS will send the RSR to indicate it has experienced a failure, and the MME or the SGSN will need to update the HSS with location information when the mobile device connects with the network, or roams (triggering a location update) (Table 6.7).

Notify-Request/Answer   When the HSS needs to know about changes to a mobile device and its associated profile, the HSS may request notification from the MME or the SGSN. For example, if there is roaming, but there is no ULR sent, the MME will send NOR to the HSS with updated information about the location of the mobile device (Table 6.8).

The MME will also send this command to notify the HSS about any changes to the PDN Gateway assignment for a specific APN. Other reasons the MME or the SGSN will send NOR include:

- There is a change in the serving node due to roaming but there is no ULR, and the HSS needs to send a CLR to the current SGSN.
- To notify the HSS that the mobile device has become reachable again.
- The mobile device has enough memory available to receive one or more short messages.

PUR	PUA
< Session-Id >	< Session-Id >
[Vendor-Specific-Application-Id]	[Vendor-Specific-Application-Id]
{Auth-Session-State}	*[Supported-Features]
{Origin-Host}	[Result-Code]
{Origin-Realm}	[Experimental-Result]
[Destination-Host]	{Auth-Session-State}
{Destination-Realm}	{Origin-Host}
{User-Name}	{Origin-Realm}
[OC-Supported-Features]	[OC-Supported-Features]
[PUR-Flags]	[OC-OLR]
*[Supported-Features]	[PUA-Flags]
[EPS-Location-Information]	*[AVP]
*[AVP]	*[Failed-AVP]
*[Proxy-Info]	*[Proxy-Info]
*[Route-Record]	*[Route-Record]

TABLE 6.6   PUR/PUA Command Content

RSR	RSA
<Session-Id>	<Session-Id>
[Vendor-Specific-Application-Id]	[Vendor-Specific-Application-Id]
{Auth-Session-State}	*[Supported-Features]
{Origin-Host}	[Result-Code]
{Origin-Realm}	[Experimental-Result]
{Destination-Host}	{Auth-Session-State}
{Destination-Realm}	{Origin-Host}
*[Supported-Features]	{Origin-Realm}
*[User-Id]	*[AVP]
*[Reset-ID]	*[Failed-AVP]
*[AVP]	*[Proxy-Info]
*[Proxy-Info]	*[Route-Record]
*[Route-Record]	

TABLE 6.7   RSR/RSA Command Content

NOR	NOA
< Session-Id >	< Session-Id >
[Vendor-Specific-Application-Id]	[Vendor-Specific-Application-Id]
{Auth-Session-State}	[Result-Code]
{Origin-Host}	[Experimental-Result]
{Origin-Realm}	{Auth-Session-State}
[Destination-Host]	{Origin-Host}
{Destination-Realm}	{Origin-Realm}
{User-Name}	[OC-Supported-Features]
[OC-Supported-Features]	[OC-OLR]
*[Supported-Features]	*[Supported-Features]
[Terminal-Information]	*[AVP]
[MIP6-Agent-Info]	*[Failed-AVP]
[Visited-Network-Identifier]	*[Proxy-Info]
[Context-Identifier]	*[Route-Record]
[Service-Selection]	
[Alert-Reason]	
[UE-SRVCC-Capability]	
[NOR-Flags]	
[Homogeneous-Support-of-IMS-Voice-Over-PS-Sessions]	
*[AVP]	
*[Proxy-Info]	
*[Route-Record]	

TABLE 6.8   NOR/NOA Command Content

Connecting to the HSS via Sh Application

The Sh application is used in IMS to connect between the application server and the HSS. The procedures are very much like those of S6a, although the commands are different. There really is not much in terms of behaviors of the AS and the HSS, at least none that this author could find.

The Diameter application identifier assigned to the Sh interface application is 16777217 (allocated by IANA).

Sh Commands

The commands used on the Sh are different from S6a, yet have many of the same functionalities. Why the 3GPP did not standardize on this interface is unclear. Following are the commands defined for the Sh interface.

User-Data-Request/Answer When the application server needs subscriber data from the HSS, it sends the User-Data-Request (UDR) command to the HSS. The HSS will then respond on Sh with the User-Data-Answer (UDA) (Table 6.9).

UDR	UDA
<Session-ID>	<Session-ID>
{Vendor-Specific-Application-ID}	{Vendor-Specific-Application-ID}
{Auth-Session-State}	[Result-Code]
{Origin-Host}	[Experimental-Result]
{Origin-Realm}	{Auth-Session-State}
[Destination-Host]	{Origin-Host}
{Destination-Realm}	{Origin-Realm}
*[Supported-Features]	*[Supported-Features]
{User-Identity}	[Wildcarded-Public-Identity]
[Wildcarded-Public-Identity]	[Wildcarded-IMPU]
[Wildcarded-IMPU]	[User-Data]
[User-Data]	[OC-Supported-Features]
*[Service-Indication]	[OC-OLR]
*{Data-Reference}	*[AVP]
*[Identity-Set]	*[Failed-AVP]
[Requested-Domain]	*[Proxy-Info]
[Current-Location]	*[Route-Record]
*[DSAI-Tag]	
[Session-Priority]	
[User-Name]	
[Requested-Nodes]	
[Serving-Node-Indication]	
[Pre-Paging-Supported]	
[Local-Time-Zone-Indication]	
[UDR-Flags]	
[Call-Reference-Info]	
[OC-Supported-Features]	
*[AVP]	
*[Proxy-Info]	
*[Route-Record]	

TABLE 6.9 UDR/UDA Command Content

SNR	SNA
<Session-ID>	<Session-ID>
{Vendor-Specific-Application-ID}	{Vendor-Specific-Application-ID}
{Auth-Session-State}	{Auth-Session-State}
{Origin-Host}	[Result-Code]
{Origin-Realm}	[Experimental-Result]
[Destination-Host]	{Origin-Host}
{Destination-Realm}	{Origin-Realm}
*[Supported-Features]	[Wildcarded-Public-Identity]
{User-Identity}	[Wildcarded-IMPU]
[Wildcarded-Public-Identity]	*[Supported-Features]
[Wildcarded-IMPU]	[User-Identity]
*[Service-Information]	[Expiry-Time]
[Send-Data-Indication]	[OC-Supported-Features]
[Server-Name]	[OC-OLR]
{Subs-Req-Type}	*[AVP]
*{Data-Reference}	*[Failed-AVP]
*[Identity-Set]	*[Proxy-Info]
[Expiry-Time]	*[Route-Record]
*[DSAI-Tag]	
[One-Time-Notification]	
[User-Name]	
[OC-Supported-Features]	
*[AVP]	
*[Proxy-Info]	
*[Route-Record]	

TABLE 6.10   SNR/SNA Command Content

Subscribe-Notifications-Request/Answer (SNR/SNA)   The application server uses SNR to subscribe to notifications from the HSS when there is a change in subscriber data. The SNA is sent by the HSS to acknowledge notification has been set (Table 6.10).

Push-Notification-Request/Answer   The application server uses the Push-Notification-Request (PNR) to request notification from the HSS when there is a change to the subscriber data. The HSS uses the Push-Notification-Answer (PNA) to send the notification (Table 6.11).

Profile-Update-Request/Answer   When the application server needs to request subscriber data from the HSS, it uses the Profile-Update-Request (PUR) to request it on the Sh interface. The HSS will then send the data in the Profile-Update-Answer (PUA) command (Table 6.12).

PNR	PNA
<Session-ID>	<Session-ID>
{Vendor-Specific-Application-ID}	{Vendor-Specific-Application-ID}
{Auth-Session-State}	[Result-code]
{Origin-Host}	[Experimental-Result]
{Origin-Realm}	{Auth-Session-State}
{Destination-Host}	{Origin-Host}
{Destination-Realm}	{Origin-Realm}
*[Supported-Features]	*[Supported-Features]
{User-Identity}	*[AVP]
[Wildcarded-Public-Identity]	*[Failed-AVP]
[Wildcarded-IMPU]	*[Proxy-Info]
[User-Name]	*[Route-Record]
{User-Data}	
*[AVP]	
*[Proxy-Info]	
*[Route-Record]	

TABLE 6.11   PNR/PNA Command Content

PUR	PUA
<Session-ID>	<Session-ID>
{Vendor-Specific-Application-ID}	{Vendor-Specific-Application-ID}
{Auth-Session-State}	[Result-Code]
{Origin-Host]	[Experimental-Result]
{Origin-Realm}	{Auth-Session-State}
[Destination-Host]	{Origin-Host}
{Destination-Realm}	{Origin-Realm}
*[Supported-Features]	[Wildcarded-Public-Identity]
{User-Identity}	[Wildcarded-IMPU]
[Wildcarded-Public-Identity]	[Repository-Data-ID]
[Wildcarded-IMPU]	[Data-Reference]
[User-Name]	*[Supported-Features]
*{Data-Reference}	[OC-Supported-Features]
{User-Data}	[OC-OLR]
[OC-Supported-Features]	*[AVP]
*[AVP]	*[Failed-AVP]
*[Proxy-Info]	*[Proxy-Info]
*[Route-Record]	*[Route-Record]

TABLE 6.12   PUR/PUA Command Content

Connecting to Equipment Identity Register via S13

The Equipment Identity Register (EIR) is a database used to store the International Mobile Equipment Identity (IMEI) of stolen devices. Prior to the device being authorized for a session, the serving node first checks the local EIR for an entry. If the IMEI appears in the database, the access will be restricted.

The network service provider populates the EIR entries manually. Since the EIR provides a view of the local network, it could have a limited effect. For this reason, the GSM Association provides a service through their IMEI database. This service allows service providers with an EIR to connect with the GSMA and share their data with all other members. This increases the value of an EIR database as now the service provider gets a more global view of stolen devices that could possibly show up in their network.

The MME or SGSN use the ME-Identity-Check-Request/Answer (ECR/ECA) command to check the database.

S13 Procedures

There are not many procedures for checking the database. It is pretty simple. The MME and SGSN know the address of the EIR as this is configured in the MME and the SGSN. When a device requests service, the MME or the SGSN (whichever is receiving the request) first sends the ECR with the IMEI of the requesting device.

When the EIR receives the ECR, it checks the IMEI to see if there is a match. If there is no match, the ECA will be returned with Result-Code AVP value of DIAMETER_ERROR_EQUIPMENT_UNKNOWN. This would mean the device has not been stolen and should not be restricted from using the network.

On the other hand, if the IMEI is in the EIR, the EIR returns the ECA with Result-Code DIAMETER_SUCCESS. Optionally the IMSI could be included in the ECA. This is an indicator that the device is stolen and should not be allowed any service.

S13 Commands

When the MME or the SGSN wish to query the EIR to see if a specific IMEI is registered as stolen, they send the ECR command. The ECR command identifies the IMEI to be checked. If the IMEI is registered in the EIR, the EIR returns the ECA with the status of the equipment.

Me-Identity-Check-Request/Answer   The Me-Identity-Check-Request (ECR) and the Me-Identity-Check-Answer (ECA) are used by both the MME and the SGSN to query the EIR (Table 6.13).

AVPs Used in This Chapter

There are many AVPs defined for the S6a/S6d interface, as well as the S13/S13' interfaces. I found it easiest to list these AVPs in alphabetical order rather than in some form of order, as many of these AVPs are reused in other parts of the protocol. You will also see where grouped AVPs contain AVPs themselves, and rather than listing those and then defining them together, I have listed these all alphabetically. This should make it easier for finding the AVP of interest and its definition.

In terms of defining the AVPs, I have provided some high-level descriptions but it is best to reference the actual 3GPP documents for the exact use of an AVP. This is

ECR	ECA
<Session-ID>	<Session-ID>
[Vendor-Specific-Application-ID]	[Vendor-Specific-Application-ID]
{Auth-Session-State}	[Result-Code]
{Origin-Host}	[Experimental-Result]
{Origin-Realm}	{Auth-Session-State}
[Destination-Host]	{Origin-Host}
{Destination-Realm}	{Origin-Realm}
{Terminal-Information}	[Equipment-Status]
[User-Name]	*[AVP]
*[AVP]	*[Failed-AVP]
*[Proxy-Info]	*[Proxy-Info]
*[Route-Record]	*[Route-Record]

TABLE **6.13** ECR/ECA Command Content

because an AVP may be used in one command for a specific application, and used differently on another interface supporting a different application. There is same AVP, but with different values and different definition.

AVPs can have a variety of values. Some are grouped AVPs, which means they will have an AVP header, and then contain several other AVPs. Each of the other AVPs contained within a grouped AVP may also be grouped, or contain specific values.

Another form found here is bit-masked AVPs. A bit-masked AVP will have a quantity of bits, and depending on whether the bit is set to a value of 1 or 0 as to the meaning of the bit. Typically, if the bit is cleared (0), it has no value. If the bit is set (1), the bit definition will be applied.

Other AVPs are enumerated. The number used will represent the value of the AVP contents. Usually the numbering starts at 0, and continues on until all the values are represented.

Table 6.14 identifies AVPs defined by 3GPP, in order of their AVP code. The AVPs are defined in alphabetical order, but the descriptions are not detailed. It is best to reference the individual standards for the interface you are working with for the complete definition.

AVP Definitions

3GPP2-MEID
The Mobile Equipment Identifier is used in non-GSM networks, such as CDMA. It is analogous to the IMEI found in GSM networks.

Access-Restriction-Data
Used to identify access services that are to be blocked for the subscription. It is sent in the Subscription-Data AVP should any of these access types be blocked for the subscriber, and is sent in the ULA command. This AVP is bit-masked, with the following definitions (Table 6.15):

Attribute Name	AVP Code	Value Type
Subscription-Data	1400	Grouped
Terminal-Information	1401	Grouped
IMEI	1402	UTF8String
Software-Version	1403	UTF8String
QoS-Subscribed	1404	OctetString
ULR-Flags	1405	Unsigned32
ULA-Flags	1406	Unsigned32
Visited-PLMN-Id	1407	OctetString
Requested-EUTRAN-Authentication-Info	1408	Grouped
Requested-UTRAN-GERAN-Authentication-Info	1409	Grouped
Number-of-Requested-Vectors	1410	Unsigned32
Re-Synchronization-Info	1411	OctetString
Immediate-Response-Preferred	1412	Unsigned32
Authentication-Info	1413	Grouped
E-UTRAN-Vector	1414	Grouped
UTRAN-Vector	1415	Grouped
GERAN-Vector	1416	Grouped
Network-Access-Mode	1417	Enumerated
HPLMN-ODB	1418	Unsigned32
Item-Number	1419	Unsigned32
Cancellation-Type	1420	Enumerated
DSR-Flags	1421	Unsigned32
DSA-Flags	1422	Unsigned32
Context-Identifier	1423	Unsigned32
Subscriber-Status	1424	Enumerated
Operator-Determined-Barring	1425	Unsigned32
Access-Restriction-Data	1426	Unsigned32
APN-OI-Replacement	1427	UTF8String
All-APN-Configurations-Included-Indicator	1428	Enumerated
APN-Configuration-Profile	1429	Grouped
APN-Configuration	1430	Grouped
EPS-Subscribed-QoS-Profile	1431	Grouped
VPLMN-Dynamic-Address-Allowed	1432	Enumerated
STN-SR	1433	OctetString
Alert-Reason	1434	Enumerate
AMBR	1435	Grouped
CSG-Subscription-Data	1436	Grouped
CSG-Id	1437	Unsigned32

TABLE 6.14 AVPs Listed by Code

Attribute Name	AVP Code	Value Type
PDN-GW-Allocation-Type	1438	Enumerated
Expiration-Date	1439	Time
RAT-Frequency-Selection-Priority-ID	1440	Unsigned32
IDA-Flags	1441	Unsigned32
PUA-Flags	1442	Unsigned32
NOR-Flags	1443	Unsigned32
User-ID	1444	UTF8String
Equipment-Status	1445	Enumerated
Regional-Subscription-Zone-Code	1446	OctetString
RAND	1447	OctetString
XRES	1448	OctetString
AUTN	1449	OctetString
KASME	1450	OctetString
Trace-Collection-Entity	1452	Address
Kc	1453	OctetString
SRES	1454	OctetString
PDN-Type	1456	Enumerated
Roaming-Restricted-due-to-Unsupported-Feature	1457	Enumerated
Trace-Data	1458	Grouped
Trace-Reference	1459	OctetString
Trace-Depth	1462	Enumerated
Trace-NE-Type-List	1463	OctetString
Trace-Interface-List	1464	OctetString
Trace-Event-List	1465	OctetString
OMC-ID	1466	OctetString
GPRS-Subscription-Data	1467	Grouped
Complete-Data-List-Included-Indicator	1468	Enumerated
PDP-Context	1469	Grouped
PDP-Type	1470	OctetString
3GPP2-MEID	1471	OctetString
Specific-APN-Info	1472	Grouped
LCS-Info	1473	Grouped
GMLC-Number	1474	OctetString
LCS-PrivacyException	1475	Grouped
SS-Code	1476	OctetString
SS-Status	1477	OctetString
Notification-To-UE-User	1478	Enumerated

TABLE 6.14 AVPs Listed by Code (*Continued*)

Attribute Name	AVP Code	Value Type
External-Client	1479	Grouped
Client-Identity	1480	OctetString
GMLC-Restriction	1481	Enumerated
PLMN-Client	1482	Enumerated
Service-Type	1483	Grouped
ServiceTypeIdentity	1484	Unsigned32
MO-LR	1485	Grouped
Teleservice-List	1486	Grouped
TS-Code	1487	OctetString
Call-Barring-Info	1488	Grouped
SGSN-Number	1489	OctetString
IDR-Flags	1490	Unsigned32
ICS-Indicator	1491	Enumerated
IMS-Voice-Over-PS-Sessions-Supported	1492	Enumerated
Homogeneous-Support-of-IMS-Voice-Over-PS-Sessions	1493	Enumerated
Last-UE-Activity-Time	1494	Time
EPS-User-State	1495	Grouped
EPS-Location-Information	1496	Grouped
MME-User-State	1497	Grouped
SGSN-User-State	1498	Grouped
User-State	1499	Enumerated
MME-Location Information	1600	Grouped
SGSN-Location-Information	1601	Grouped
E-UTRAN-Cell-Global-Identity	1602	OctetString
Tracking-Area-Identity	1603	OctetString
Cell-Global-Identity	1604	OctetString
Routing-Area-Identity	1605	OctetString
Location-Area-Identity	1606	OctetString
Service-Area-Identity	1607	OctetString
Geographical-Information	1608	OctetString
Geodetic-Information	1609	OctetString
Current-Location-Retrieved	1610	Enumerated
Age-Of-Location-Information	1611	Unsigned32
Active-APN	1612	Grouped
Error-Diagnostic	1614	Enumerated
Ext-PDP-Address	1621	Address
UE-SRVCC-Capability	1615	Enumerated

TABLE 6.14   AVPs Listed by Code (*Continued*)

Attribute Name	AVP Code	Value Type
MPS-Priority	1616	Unsigned32
VPLMN-LIPA-Allowed	1617	Enumerated
LIPA-Permission	1618	Enumerated
Subscribed-Periodic-RAU-TAU-Timer	1619	Unsigned32
Ext-PDP-Type	1620	OctetString
SIPTO-Permission	1613	Enumerated
MDT-Configuration	1622	Grouped
Job-Type	1623	Enumerated
Area-Scope	1624	Grouped
List-Of-Measurements	1625	Unsigned32
Reporting-Trigger	1626	Unsigned32
Report-Interval	1627	Enumerated
Report-Amount	1628	Enumerated
Event-Threshold-RSRP	1629	Unsigned32
Event-Threshold-RSRQ	1630	Unsigned32
Logging-Interval	1631	Enumerated
Logging-Duration	1632	Enumerated
Relay-Node-Indicator	1633	Enumerated
MDT-User-Consent	1634	Enumerated
PUR-Flags	1635	Unsigned32
Subscribed-VSRVCC	1636	Enumerated
Equivalent-PLMN-List	1637	Grouped
CLR-Flags	1638	Unsigned32
UVR-Flags	1639	Unsigned32
UVA-Flags	1640	Unsigned32
VPLMN-CSG-Subscription-Data	1641	Grouped
Time-Zone	1642	UTF8String
A-MSISDN	1643	OctetString
MME-Number-for-MT-SMS	1645	OctetString
SMS-Register-Request	1648	Enumerated
Local-Time-Zone	1649	Grouped
Daylight-Saving-Time	1650	Enumerated
Subscription-Data-Flags	1654	Unsigned32
Measurement-Period-UMTS	1655	Enumerated
Measurement-Period-LTE	1656	Enumerated
Collection-Period-RRM-LTE	1657	Enumerated
Collection-Period-RRM-UMTS	1658	Enumerated

TABLE 6.14 AVPs Listed by Code (*Continued*)

Attribute Name	AVP Code	Value Type
Positioning-Method	1659	OctetString
Measurement-Quantity	1660	OctetString
Event-Threshold-Event-1F	1661	Integer32
Event-Threshold-Event-1I	1662	Integer32
Restoration-Priority	1663	Unsigned32
SGs-MME-Identity	1664	UTF8String
SIPTO-Local-Network-Permission	1665	Unsigned32
Coupled-Node-Diameter-ID	1666	DiameterIdentity
WLAN-offloadability	1667	Grouped
WLAN-offloadability-EUTRAN	1668	Unsigned32
WLAN-offloadability-UTRAN	1669	Unsigned32
Reset-ID	1670	OctetString
MDT-Allowed-PLMN-Id	1671	OctetString

TABLE 6.14   AVPs Listed by Code (*Continued*)

Bit	Description
0	UTRAN not allowed
1	GERAN not allowed
2	GAN not allowed
3	I-HSPA-Evolution not allowed
4	E-UTRAN not allowed
5	HO to non-3GPP access not allowed

TABLE 6.15   Bit Values for Access-Restriction-Data AVP

Active-APN

This grouped AVP provides information about an APN that was established dynamically. The HSS uses this information in the event of a node restart.

<Active-APN>::=<AVP Header: 1612, 10415>

{Context-Identifier}

[Service-Selection]

[MIP6-Agent-Info]

[Visited-Network-Identifier]

*[Specific-APN-Info]

*[AVP]

Age-of-Location-Information
This AVP contains the elapsed time in minutes since the last radio contact with the mobile device.

Alert-Reason
This AVP is enumerated with two values.

 0 = Mobile device present

 1 = Mobile device memory available

All-APN-Configurations-Included-Indicator
When the MME or the SGSN receive an APN-Configuration-Profile AVP, they check first to see if there is a match to an existing APN configuration. If there is a match, the existing APN configuration is replaced with the new information. If there is no match, the received APN configuration will be considered as new, and will be added to the subscription.

 If the value of this AVP is MODIFIED_ADDED_APN_CONFIGURATIONS_ INCLUDED, the MME or SGSN will check the Context-Identifier in the APN configuration for a match.

 This AVP is enumerated with the following possible values:

 0 = ALL_APN_CONFIGURATIONS_INCLUDED

 1 = MODIFIED_ADDED_APN_CONFIGURATIONS_INCLUDED

Allocation-Retention-Priority
This AVP indicates the priority and length of retention for the identified APN. This is a grouped AVP.

 <Allocation-Retention-Priority>::=<AVP Header: 1034, 10415>

 {Priority-Level}

 [Pre-emption-Capability]

 [Pre-emption-Vulnerability]

The default value is considered to be disabled if the Pre-emption-Capability AVP is absent, meaning pre-emption will not apply. The default is considered to be enabled if the Pre-emption-Vulnerability is absent.

AMBR
This is a grouped AVP used to indicate the Aggregate Maximum Bit Rate to be applied.

 <AMBR>::=<AVP Header: 1435, 10415>

 {Max-Requested-Bandwidth-UL}

 {Max-Requested-Bandwidth-DL}

 *[AVP]

A-MSISDN

This is used to communicate an Additional MSISDN (A-MSISDN). It is formatted according to ITU E.164, but there are no indicators identifying the nature of address or the numbering plan as seen in MAP. This AVP is used in the Subscription-Data AVP sent in the ULA, as well as in an IDR.

APN-Configuration

Each subscription has a unique APN-Configuration identified by the Context-Identifier. This AVP provides the details for each APN-Configuration. The Served-Party-IP-Address can be present up to two times, and carries the static IPv4 or IPv6 address of the device. The Visited-Network-Identifier indicates the address of the network where the packet data network gateway is allocated, but only when dynamic gateway assignment is being used.

<APN-Configuration>::=<AVP Header: 1430, 10415>

{Context-Identifier}

*2 [Served-Party-IP-Address]

{PDN-Type}

{Service-Selection}

[EPS-Subscribed-QoS-Profile]

[VPLMN-Dynamic-Address-Allowed]

[MIP6-Agent-Info]

[Visited-Network-Identifier]

[PDN-GW-Allocation-Type]

[3GPP-Charging-Characteristics]

[AMBR]

*[Specific-APN-Info]

[APN-OI-Replacement]

[SIPTO-Permission]

[LIPA-Permission]

[Restoration-Priority]

[SIPTO-Local-Network-Permission]

[WLAN-Offloadability]

*[AVP]

APN-Configuration-Profile

Contains information about the subscribers APN configurations in the packet network. The default APN configuration is identified in the Context-Identifier AVP. The Subscription-Data AVP will contain one APN-Configuration-Profile for each IMSI. Each APN-Configuration-Profile contains one or more APN-Configurations. Each APN-Configuration provides the configuration for a single APN.

<APN-Configuration-Profile>::=<AVP Header: 1429, 10415>

{Context-Identifier}

{All-APN-Configurations-Included-Indicator}

1*{APN-Configuration}

*[AVP]

APN-OI-Replacement

This AVP is used to provide the domain name that replaces the APN OI in a non-roaming case. In the case of home routing, the fully qualified domain name (FQDN) is provided to support DNS resolution.

It is sent in the Subscription-Data AVP in an IDR command. The format for the value is a character string.

Area-Scope

This is a grouped AVP identifying the geographical area where measurements should be taken when minimalized drive tests (MDT) is being used. The geographical area is expressed in terms of cell IDs or other network identities.

<Area-Scope>::=<AVP Header: 1623, 10415>

[Cell-Global-Identity]

*[E-UTRAN-Cell-Global-Identity]

*[Routing-Area-Identity]

*[Location-Area-Identity]

*[Tracking-Area-Identity]

*[AVP]

AS-Number

This AVP is used on the Sh interface to provide the application server number in the IMS.

Authentication-Info

This is a grouped AVP used by the HSS to provide authentication information to the requesting node.

< Authentication-Info>::=<AVP Header: 1413, 10415>

*[E-UTRAN-Vector]

*[UTRAN-Vector]

*[GERAN-Vector]

*[AVP]

AUTN

This AVP is used as part of authentication to provide authentication vectors.

Call-Barring-Info

This AVP contains the service codes for call barring services related to short message service for a subscriber.

<Call-Barring-Info>::=<AVP Header: 1488, 10415>

{SS-Code}

{SS-Status}

*[AVP]

Call-Reference-Info

This is a grouped AVP used on the Sh interface. It contains the call reference number and application server number.

<Call-Reference-Info>::=<AVP Header; 720, 10415>

{Call-Reference-Number}

{AS-Number}

*[AVP]

Call-Reference-Number

This AVP is used on the Sh interface, and provides the call reference number assigned by the call control. This is represented as an octet string.

Cancellation-Type

This AVP is used by the HSS to designate the reason for sending the Cancel-Location command. The following enumerated values are supported:

0 = MME update procedure—this is sent when the HSS has received a Update-Location-Request from another MME indicating the subscriber has roamed to another serving MME.

1 = SGSN update procedure—this is sent when the HSS has received a Update-Location-Request from another SGSN indicating the subscriber has roamed to another serving SGSN.

2 = Subscription withdrawal—this is sent by the HSS to the MME or SGSN when the HSS administrator has withdrawn a subscription from the HSS.

3 = Update procedure IWF—used when interworking with a Rel-8 HSS.

4 = Initial detach procedure—sent by the HSS when an Update-Location-Request has been received as part of an initial attach from another MME or SGSN.

Cell-Global-Identity

This is used to send the cell global identification where the subscriber is currently registered. The cell identifier is comprised of the MCC and the MNC.

Client-Identity

This AVP contains the ISDN number of the external client that is allowed to retrieve location information for a mobile device. The value is in E.164 format.

Bit	Description
0	Indicates the CLR command was sent on S6a when the value is set to 1. If the value is set to 0, the command was sent on the S6d interface.
1	Indicates the MME/SGSN will request the mobile device to initiate immediate re-attach procedure when set to the value of 1.

TABLE 6.16 Bit Values for CLR-Flags AVP

CLR-Flags

The CLR flags are sent during the initial attach for a combined MME/SGSN. It is a bit-masked AVP with the following bits set to the value of either 1 or 0 (Table 6.16).

Collection-Period-RRM-LTE

This AVP is used when MDT is being implemented. It is only used when measurements are being collected in the radio access network (RAN) in an LTE network. The AVP defines the period to be applied for measurements collection in milliseconds, up to one minute. It is mandatory if the job type is "immediate MDT" or "immediate MDT and trace," and any of the bits (2 or 3) in the List-of-Measurements AVP are set to one. The values are enumerated, with the following potential values:

0 = 1024 ms

1 = 1280 ms

2 = 2048 ms

3 = 2560 ms

4 = 5120 ms

5 = 10,240 ms

6 = 1 minute

Collection-Period-RRM-UMTS

This AVP is used when MDT is being implemented. It is only used when measurements are being collected in the radio access network (RAN) in an UMTS network. The AVP defines the period to be applied for measurements collection in milliseconds, up to one minute. It is mandatory if the job type equals "immediate MDT" or "immediate MDT and trace," and any of the bits 3, 4, or 5 in the List-of-Measurements AVP are set to the value of 1. The values are enumerated, with the following potential values:

0 = 250 ms

1 = 500 ms

2 = 1000 ms

3 = 2000 ms

4 = 3000 ms

5 = 4000 ms

6 = 6000 ms

7 = 8000 ms

8 = 12,000 ms

9 = 16,000 ms

10 = 20,000 ms

11 = 24,000 ms

12 = 28,000 ms

13 = 32,000 ms

14 = 64,000 ms

Complete-Data-List-Included-Indicator
This AVP is enumerated, with only two possible values:

0 = All PDP contexts included

1 = Modified added PDP contexts included.

Context-Identifier
The Context-Identifier can be found in the APN-Configuration-Profile AVP sent in the Subscription-Data AVP. These are all sent in the ULA command. The APN-Configuration-Data AVP contains information about the subscribers APN configurations, and the Context-Identifier is used to provide a unique identifier for the default APN configuration.

Coupled-Node-Diameter-ID
When the MME and the SGSN is a combined node, the combined node uses this AVP to provide its identity. This allows the HSS to determine if a mobile device is being served by the same MME and SGSN, or by separate nodes that are not combined.

When the MME and the SGSN are a combined node, the HSS sends a single subscription update message (IDR or DSR) when applicable. The value of this AVP will be the Diameter identity of the combined node.

CSG-ID
The CSG-ID identifies a closed subscriber group (CSG). A CSG is a group of cells in 3G or 4G networks that are open to a certain group of subscribers. It is part of the CSG-Subscription-Data AVP, which provides additional information about this group such as which APNs can be accessed via this group. This AVP identifies the specific identity assigned to the CSG. It is a fixed length of 27 bits.

CSG-Subscription-Data
The CSG subscription data identifies the APNs that can be accessed by the group, as well as other information about the specific group.

<CSG-Subscription-Data>::=<AVP Header: 1436, 10415>

{CSG-ID}

[Expiration-Data]

*[Service-Selection]

[Visited-PLMN-ID]

*[AVP]

Current-Location

This AVP is used on the Sh interface to indicate if an active location retrieval is needed. The possible enumerated values are:

0 = Do not need to initiate an active location retrieval

1 = Initiate an active location retrieval

Current-Location-Retrieved

This AVP is enumerated and is used to indicate that a mobile device location was retrieved either by paging the device while it was in idle mode, or by retrieving from the eNB when the device was in connected mode. There is only one value:

0 = Active location retrieval.

Data-Reference

The value contained in this AVP is enumerated, and indicates the type of user data being requested. This is used both in the UDR and the SNR commands. Following are the possible values:

0 = Repository data

10—IMS public identity

11—IMS user state

12—S-CSCF name

13—Initial filter criteria

14—Location information

15—User state

16—Charging information

17—MSISDN

18—PSI activation

19—DSAI

20—Reserved

21—Service level trace info

22—IP address secure binding information

23—Service priority level

24—SMS registration info

25—UE reachability for IP

26—TADS information

27—STN-SR

28—UE SRVCC capability

29—Extended priority

30—CSRN

31—Reference location information

32—IMSI

33—IMS private user identity

Daylight-Saving-Time

Indicates if there is to be a time zone adjustment on the mobile device wherever it is located. This AVP is enumerated with the following values:

0 = No adjustment

1 = Plus one hour adjustment

2 = Plus two hour adjustment

DSA-Flags

This AVP is bit-masked. When the bit is set to the value of 1, the bit is considered as set and has the meaning shown in Table 6.17.

DSAI-Tag

The dynamic service activation information tag identifies what is being accessed for the public identity. It is sent as an octet string.

DSR-Flags

This AVP is bit-masked, with each bit representing a specific value. When the value of the bit is 0, the bit is considered as cleared and is ignored. When the bit value is 1, the bit is considered set and its corresponding definition is considered during processing of the message. The values can be found in Table 6.18:

EPS-Location-Information

This AVP contains information about the location of the mobile device. It is a grouped AVP.

<EPS-Location-Information>::=<AVP Header: 1496, 10415>

[MME-Location-Information]

[SGSN-Location-Information]

*[AVP]

Bit	Value when Set
0	Network node area restricted. Indicates that the SGSN area is restricted because of a regional subscription.

TABLE **6.17** Bit Values for DSA-Flags AVP

Bit	Value when Set
0	Regional subscription withdrawal. Indicates that any regional subscription data will be removed.
1	Complete APN configuration profile withdrawal. Indicates that all the packet APN configuration data will be deleted from the subscriber data.
2	Subscribed charging characteristics withdrawal. Indicates that charging characteristics will be deleted from the subscription data.
3	PDN subscription contexts withdrawal. Indicates that the PDN subscription contexts will be deleted from the subscription data. The identifiers for the PDN contexts are included in the Context-Identifier AVP.
4	STN-SR. Indicates the session transfer number for SRVCC will be deleted from the subscription data.
5	Complete PDP context list withdrawal. Indicates that the PDP contexts will be deleted from the subscriber data.
6	PDP context withdrawals. Used to indicate the identified PDP contexts will be deleted from the subscriber data.
7	Roaming restricted due to unsupported feature. Used to indicate a roaming restriction will be removed from the subscription data.
8	Trace data withdrawal. Used to indicate trace data will be removed from the subscriber data
9	CSG deleted. Used to indicate the CSG-Subscription-Data will be deleted from the subscription data
10	APN-OI-Replacement. Used to indicate the device level APN-OI-Replacement will be deleted
11	GMLC list withdrawal. Indicates the list of the subscribers GMLCs will be deleted
12	LCS withdrawal. Indicates the LCS service identified in the SS-Code AVP will be removed
13	SMS withdrawal. Indicates the SMS service identified in the SS-Code AVP will be deleted
14	Subscribed periodic RAU/TAU timer withdrawal. Indicates the subscribed periodic RAU/TAU timer value will be deleted
15	Subscribed VSRVCC withdrawal. Indicates the subscribed VSRVCC will be deleted
16	A-MSISDN withdrawal. Indicates the additional MSISDN shall be deleted
17	ProSe withdrawal. Indicates Proximity Services data will be deleted
18	Reset-IDs. Indicates the set of Reset-IDs will be removed

TABLE 6.18   Bit Values for DSR-Flags AVP

EPS-Subscribed-QoS-Profile

This AVP carries the QoS parameters assigned for the bearer channel for a given APN.

<EPS-Subscribed-QoS-Profile>::=<AVP Header: 1431, 10415>

{QoS-Class-Identifier}

{Allocation-Retention-Priority}

*[AVP]

EPS-User-State

This AVP contains information about the state of the user in the MME or the SGSN. This is a grouped AVP.

<EPS-User-State>::=<AVP Header: 1495, 10415>

[MME-User-State]

[SGSN-User-State]

*[AVP]

Equipment-Status

This AVP indicates the status of the device in the EIR. It is only used on the S13/S13' interfaces. The values are enumerated:

0 = Device is whitelisted

1 = Device is blacklisted

2 = Device is greylisted

Equivalent-PLMN-List

Contains the identifiers for all registered networks where the MME or SGSN reside. This is a grouped AVP.

<Equivalent-PLMN-List>::=<AVP Header: 1637, 10415>

1*{Visited-PLMN-ID}

*[AVP]

Error-Diagnostic

This is an enumerated AVP that provides information about error messages sent. The following values are supported:

0 = GPRS data subscribed. This is used when the Experimental-Error AVP is set to DIAMETER_ERROR_UNKNOWN_EPS_SUBSCRIPTION is sent and there is GPRS subscription data for the user.

1 = No GPRS data subscribed. Used when the Experimental-Error AVP is set to the value of DIAMETER_ERROR_UNKNOWN_EPS_SUBSCRIBER is sent and there is no GPRS subscription data for the user.

2 = Operator Determined Barring (ODB) all APNs. Used when the Experimental-Error AVP is set to DIAMETER_ERROR_ROAMING_NOT_ALLOWED and operator determined barring is set to "all packet-oriented services are barred."

3 = Operator Determined Barring (ODB) VPLMN APN. This is used when the Experimental-Error value is DIAMETER_ERROR_ROAMING_NOT_ALLOWED and operator determined barring is set to "roamer access HPLMN access point barred."

4 = Operator Determined Barring (ODB) VPLMN APN. This is used when the Experimental-Error value is set to DIAMETER_ERROR_ROAMING_NOT_ALLOWED and operator determined barring is set to "roamer access to VPLMN-AP barred."

MCC	MNC	LAC	CI
Location Area Identification			
Cell Global Identification (CGI)			

TABLE 6.19 *AVP Structure for E-UTRAN-Cell-Global-Identity AVP*

E-UTRAN-Cell-Global-Identity

This AVP identifies the cell in which the mobile device is currently registered. The format follows the MAP protocol where the cell ID is identified by the mobile country code and the mobile network code. The cell global identity is the combined LAI and the CI (Table 6.19).

E-UTRAN-Vector

Authentication credentials are provided by the HSS to the MME when authentication is requested. This is a grouped AVP.

> < E-UTRAN-Vector>::=<AVP Header: 1414, 10415>
>
> [Item-Number]
>
> {RAND}
>
> {XRES}
>
> {AUTN}
>
> {KASME}
>
> *[AVP]

Event-Threshold-Event-1F

This AVP is used with MDT when the job type is equal to "immediate MDT" or "immediate MDT and trace," and the reporting trigger is equal to 1F event reporting. It defines the reporting threshold for UMTS M1 measurements. The value is a number between 120 and 165. The range is dependent on the measurement:

- CPICH RSCP range = −120 to 25 dBm
- CPICH Ec/No range = −24 to 0 dB
- Path loss = 30 to 165 dB

Event-Threshold-Event-1I

This AVP is also used for MDT procedures and is mandatory if the job type equals "immediate MDT" or " immediate MDT and trace," and the reporting trigger is equal to 1I event reporting.

The AVP defines the reporting threshold for UMTS M2 measurements for 1I event-based reporting. The value is a number between −120 and −25.

Event-Threshold-RSRP

This AVP is used for MDT procedures and is mandatory when the job type is equal to "immediate MDT" or "immediate MDT and trace," and the reporting trigger is configured for A2 event reporting or A2 event triggered periodic reporting.

The value is a number between 0 and 97.

Event-Threshold-RSRQ

This AVP is used for MDT procedures and is mandatory if the job type is equal to "immediate MDT" or "immediate MDT and trace," and the report trigger is configured for A2 event reporting or A2 event triggered periodic reporting.

The value provided in this AVP is a number between 0 and 34.

Experimental-Result

Each application will have unique error codes defined by the vendor (in this case, the vendor is 3GPP). The Result-Code AVP will communicate success or errors based on the base protocol (see Chap. 2 for Result-Code AVP values). For the S6a/S6d application, the following error codes are supported in this AVP:

Transient Failures

DIAMETER_AUTHENTICATION_DATA_AVAILABLE (4181)

This is sent by the HSS to indicate a transient failure occurred and the requesting node should attempt its request again later.

DIAMETER_ERROR_CAMEL_SUBSCRIPTION_PRESENT (4182)

This code is sent by the HSS to indicate the subscriber being registered has SGSN CAMEL subscription data.

Permanent Failures

DIAMETER_ERROR_USER_UNKNOWN (5001)

This is used by the HSS to indicate the IMSI received in a command is unknown.

DIAMETER_ERROR_UNKNOWN_EPS (5420)

Sent by the HSS to indicate there is no EPS subscription for the received IMSI.

DIAMETER_ERROR_RAT_NOT_ALLOWED (5421)

This is sent by the HSS to indicate the RAT being used by the mobile device is not allowed for the received IMSI.

DIAMETER_ERROR_ROAMING_NOT_ALLOWED (5004)

This is sent by the HSS to indicate the subscriber is not allowed to roam within the MME or SGSN service area.

DIAMETER_ERROR_EQUIPMENT_UNKNOWN (5442)

This is sent by the EIR to indicate the mobile equipment is not known by the EIR.

DIAMETER_ERROR_UNKNOWN_SERVING_NODE (5423)

This is sent by the HSS to indicate a Notify command has been received from a node that is not currently registered as serving this subscriber.

Expiration-Date

This AVP indicates the time when subscription to a CSG-ID shall expire, expressed in date and time format.

Expiry-Time

This AVP contains the time of expiration for subscription notifications in the HSS. The value is formatted in time and date.

External-Client

This AVP identifies the external clients that are allowed to local target mobile devices for an MT-LR (mobile terminating location request).

<External-Client>::=<AVP Header: 1479, 10415>

{Client-Identity}

[GMLC-Restriction]

[Notification-to-UE-User]

*[AVP]

Ext-PDP-Address

This AVP is only present when the PDP-Address AVP is used, and contains an additional IP address. This usually happens when the PDP supports both IPv4 and IPv6. It cannot contain the same type of IP address as the PDP-Address AVP. The format is an IP address.

Ext-PDP-Type

This AVP is used with PDP-Type to indicate the support of both IPv4 and IPv6 as a dual stack. The value is expressed in hexadecimal as 8D.

Feature-List

This AVP is bit-masked. Each bit can be set to a value of 0 or 1. There are two lists of features supported (identified in the Feature-List-ID AVP). The features for Feature-List-ID #1 are given in Table 6.20.

The features for Feature-List-ID #2 are given in Table 6.21.

Geodetic-Information

This AVP is used to provide the geodetic location of the mobile device. It includes the screening and presentation indicators, type of shape (ellipsoid point with uncertainty circle), degrees of latitude, longitude, uncertainty code, and confidence indicators as an octet string.

Geographical-Information

Similar to the geodetic information, this AVP provides the geographic information for a mobile device. It is an octet string that includes the type of shape (ellipsoid point with uncertainty circle), degrees of latitude and longitude, and uncertainty code.

GERAN-Vector

This AVP is used by the HSS to return authentication credentials to the requesting SGSN. This is a grouped AVP.

< GERAN-Vector>::=<AVP Header: 1416, 10415>

[Item-Number]

{RAND}

{SRES}

{Kc}

*[AVP]

Bit	Description
0	ODB all APN. Operator Determined Barring is set for all packet services.
1	ODB HPLMN APN. Operator Determined Barring is set for packet services from access points within the home network while the subscriber is roaming. Sent in the ULR/ULA or IDR/IDA commands. The HSS would reject an Update-Location-Request (ULR) by sending ULA with Result-Code of DIAMETER_ERROR_ROAMING_NOT_ALLOWED.
2	ODB VPLMN APN. Operator Determined Barring is set for all packet services from access points within the visited network. The HSS would reject an Update-Location-Request (ULR) by sending ULA with Result-Code of DIAMETER_ERROR_ROAMING_NOT_ALLOWED.
3	ODB all outgoing calls. Operator Determined Barring is set for all outgoing calls. The HSS would reject an Update-Location-Request (ULR) by sending ULA with Result-Code of DIAMETER_ERROR_ROAMING_NOT_ALLOWED.
4	ODB all international outgoing calls. Operator Determined Barring is set for all outgoing International calls. The HSS would reject an Update-Location-Request (ULR) by sending ULA with Result-Code of DIAMETER_ERROR_ROAMING_NOT_ALLOWED.
5	ODB all international outgoing calls not to the home network. Operator Determined Barring is set for all international outgoing calls that are not directed to the home network. The HSS would reject an Update-Location-Request (ULR) by sending ULA with Result-Code of DIAMETER_ERROR_ROAMING_NOT_ALLOWED.
6	ODB all interzonal outgoing calls. Operator Determined Barring is set to block all interzonal outgoing calls. The HSS would reject an Update-Location-Request (ULR) by sending ULA with Result-Code of DIAMETER_ERROR_ROAMING_NOT_ALLOWED.
7	ODB all interzonal outgoing calls not to the home country. Operator Determined Barring is set to block all outgoing interzonal calls that are not directed to the home country. The HSS would reject an Update-Location-Request (ULR) by sending ULA with Result-Code of DIAMETER_ERROR_ROAMING_NOT_ALLOWED.
8	ODB all interzonal outgoing calls and international outgoing calls not to the home country. Operator Determined Barring is set to block all outgoing interzonal and international calls not directed to the home country. The HSS would reject an Update-Location-Request (ULR) by sending ULA with Result-Code of DIAMETER_ERROR_ROAMING_NOT_ALLOWED.
9	Regional subscriber. The HSS would reject an Update-Location-Request (ULR) by sending ULA with Result-Code of DIAMETER_ERROR_ROAMING_NOT_ALLOWED.
10	Trace. If the MME or SGSN do not support the trace feature, the HSS will not include trace data in an ULR or IDR. If the HSS has included trace data in the ULR or IDR, but the MME or SGSN do not support the feature, the HSS simply stores the data.
11	LCS all privacy exception classes. This is used by the SGSN on the S6d interface if the SGSN is also supporting the SS7 MAP-based Lg interface. If the SGSN does not support this feature, the HSS will not send the LCS information via the ULA.
12	LCS universal. Also applies to the SGSN on the S6d interface if the SGSN is supporting the MAP-based Lg interface. If the SGSN does not support this feature, the HSS will not send the LCS information via the ULA.
13	LCS call session related. Allow location by any value-added client to which a call/session is established from the target device. Only used by an SGSN also supporting the MAP-based Lg interface. If the SGSN does not support this feature, the HSS will not send the LCS information via the ULA.

TABLE 6.20   Bit Values for Feature-List AVP

Bit	Description
14	LCS call session unrelated. Allow location by designated external value-added LCS clients. Only used by an SGSN also supporting the MAP-based Lg interface. If the SGSN does not support this feature, the HSS will not send the LCS information via the ULA.
15	LCS PLMN operator. Allow location by designated PLMN operator LCS clients. Only used when the SGSN also supports the MAP-based Lg interface. If the SGSN does not support this feature, the HSS will not send the LCS information via the ULA.
16	LCS service type. Allow location by LCS clients of a designated LCS service type. Only used when the SGSN also supports the MAP-based Lg interface. If the SGSN does not support this feature, the HSS will not send the LCS information via the ULA.
17	LCS all mobile originated location request classes. Only used when the SGSN also supports the MAP-based Lg interface. If the SGSN does not support this feature, the HSS will not send the LCS information via the ULA.
18	LCS basic self-location. Allows a mobile device to request its own location data. Only used when the SGSN also supports the MAP-based Lg interface. If the SGSN does not support this feature, the HSS will not send the LCS information via the ULA.
19	LCS autonomous self-location. Allows a mobile device to perform self-location without interaction with the network. Only used when the SGSN also supports the MAP-based Lg interface. If the SGSN does not support this feature, the HSS will not send the LCS information via the ULA.
20	LCS transfer to third party. Allows a mobile device to request transfer of its location to another LCS client. Only used when the SGSN also supports the MAP-based Lg interface. If the SGSN does not support this feature, the HSS will not send the LCS information via the ULA.
21	Short message mobile originated point to point. Sent by the HSS, but if the MME or the SGSN does not support this feature, the HSS will not send this information in the ULA.
22	Barring outgoing calls. Sent by the HSS with associated SMS information if the MME or SGSN indicates support for this feature.
23	Barring all outgoing calls. Sent by the HSS with associated SMS information if the MME or SGSN indicates support for this feature.
24	Barring of outgoing international calls. Sent by the HSS with associated SMS information if the MME or SGSN indicates support for this feature.
25	Barring of outgoing international calls except those destined for the subscribers home country. Sent by the HSS with associated SMS information if the MME or SGSN indicates support for this feature.
26	Device reachability notification. The HSS will set the "UE-Reachability-Request bit in the IDR-Flags AVP and send to the MME or the SGSN if either the MME or SGSN indicates it supports the feature.
27	Terminating access domain selection data retrieval. When the MME or the SGSN indicates it supports this feature, the HSS sets the T-ADS-Data-Request bit in the IDR-Flags AVP when sending an IDR or IDA.
28	State and location information retrieval. If the MME or the SGSN support this feature, the HSS sets the EPS-User-State-Request, EPS-Location-Information-Request, or Current-Location-Request bits in the IDR-Flags AVP.

TABLE 6.20 Bit Values for Feature-List AVP (*Continued*)

Bit	Description
29	Partial purge. If the HSS indicates in ULA that it does not support this feature, the MME or the SGSN will not indicate the serving node where purge has been implemented.
30	Local time zone retrieval. If the MME or the SGSN indicate they support this feature, the Local-Time-Zone-Request bits in the IDR-Flags are set by the HSS when sending an IDR.
31	Additional MSISDN. The HSS populates the A-MSISDN AVP with the assigned MSISDN or additional MSISDN, based on the network policy, if the MME or SGSN support this feature.

TABLE 6.20   Bit Values for Feature-List AVP (*Continued*)

GMLC-Number

This AVP provides the ISDN number of the GMLC using the numbering format defined by E.164. There are no leading indicators to provide nature of address or numbering plan, only the GMLC number.

GMLC-Restriction

This can be found in the External-Client AVP. This AVP is enumerated, with the following values:

0 = GMLC list

1 = Home Country

GPRS-Subscription-Data

This AVP contains the subscription information related to GPRS.

<GPRS-Subscription-Data>::=<AVP Header: 1467, 10415>

{Complete-Data-List-Included-Indicator}

1*50 {PDP-Context}

*[AVP]

Homogeneous-Support-of-IMS-Voice-Over-PS-Sessions

This AVP is simply indicating if this feature is supported. There are only two possible enumerated values:

0 = not supported

1 = supported

HPLMN-ODB

The AVP identifies the services that are barred for this subscription. This AVP uses bit-masking. If the value of a bit is 0, the bit is considered as cleared. If the value of a bit is 1, the bit is considered "set" (Table 6.22).

Bit	Description
0	SMS in MME. Notifies the HSS that the MME can transfer SMS without establishing a signaling gateway association with the MSC. Sent in the ULR/ULA, IDR/IDA, DSR/DSA, and/or NOR/NOA commands.
1	SMS in SGSN. The HSS indicates SMS-in-SGSN-Allowed in the ULA on S6d if the SGSN supports this feature.
2	Diameter LCS all privacy exception classes. If the SGSN supports the Diameter-based Lgd interface, LCS information is sent in ULR/ULA as well as IDR/IDA commands over the S6d interface.
3	Diameter LCS universal. Allows location requests by any client if the SGSN supports the Diameter-based Lgd interface. The HSS sends the location data in the ULA over the S6d interface.
4	Diameter LCS call session related. Allows location requests of a mobile to where a call or session has been established by any client, if the SGSN supports the Diameter-based Lgd interface. The HSS sends the location data in the ULA over the S6d interface.
5	Diameter LCS call session unrelated. Allows location requests of a mobile to where a call or session has not been established by any client, if the SGSN supports the Diameter-based Lgd interface. The HSS sends the location data in the ULA over the S6d interface.
6	Diameter LCS PLMN operator. Allows location requests by designated PLMN operator location clients.
7	Diameter LCS service type. Allows location requests by LCS clients of a designated location services type.
8	Diameter LCS all mobile originated location request classes. Used when the SGSN supports the Diameter-based Lgd interface.
9	Diameter LCS basic self-location. Allows a mobile device to request its own location. Used if the SGSN supports the Diameter-based Lgd interface.
10	Diameter LCS autonomous self-location. Allows a mobile device to perform self-location without any network interaction if supported. Used if the SGSN supports the Diameter-based Lgd interface.
11	Diameter LCS transfers to third party. Allows a mobile device to request transfer of its location to another LCS client if supported. Used if the SGSN supports the Diameter-based Lgd interface.
12	Diameter Gdd in SGSN. Means the SGSN supports the Diameter-based interface Gdd, for SMS in the SGSN.
13	Optimized LCS procedural support. Means the MME and the SGSN support optimized LCS procedures.
14	SGSN CAMEL capability. Indicates the SGSN supports CAMEL.
15	Proximity Services (ProSe) capability. Indicates the MME supports proximity services.
16	P-CSCF restoration. Indicates the MME or the SGSN support execution of P-CSCF restoration.
17	Reset IDs. HSS will send the Reset-IDs AVP in the ULR/ULA, IDR/IDA, DSR/DSA, or RSR/RSA commands to the MME or the SGSN if this is supported.

TABLE **6.21** Bit Values for Feature-List AVP #2

Bit	Value if Set
0	HPLMN specific barring type 1
1	HPLMN specific barring type 2
2	HPLMN specific barring type 3
3	HPLMN specific barring type 4

TABLE 6.22   Bit Values for HPLMN-ODB AVP

ICS-Indicator

This AVP indicates whether or not IMS centralized services (ICS) are to be supported for the mobile device. ICS allows services to be provided and controlled through the IMS, regardless if the access type is circuit or packet switched. This is an enumerated AVP with the following values:

0 = False

1 = True

IDA-Flags

This AVP consists of one bit. If the bit is set to the value of 1, the SGSN area is restricted because of a regional subscription. This means the device is limited to services within the SGSN coverage area.

Identity-Set

This AVP is used on the Sh interface to indicate which public identities are being requested from the HSS. The following values are enumerated:

0 = All identities

1 = Registered identities

2 = Implicit identities

3 = Alias identities

IDR-Flags

This AVP is bit-masked. Each bit, if set to the value of 1, has the following definition (Table 6.23):

IMEI

This AVP contains the 14-digit International Mobile Equipment Identity (IMEI) used in GSM networks to identify the type of mobile equipment being used. The IMEI consists of an 8-digit type allocation code (TAC) and a 6-digit serial number. In some cases a 15th digit can be found, but is sometimes ignored by the receiving node.

Immediate-Response-Preferred

The simple presence of this AVP indicates an immediate response is requested for authentication vectors. The HSS may use the value to determine the number of vectors that are required for authentication by the MME or the SGSN.

Bit	Description
0	UE reachability request. Indicates to the MME or the SGSN that the HSS is waiting for notification of the mobile device reachability.
1	T-ADS data request. This bit indicates the HSS is requesting the support status of IMS voice over packet service. It also indicates the radio access type (RAT) and the timestamp for the last radio contact with the mobile device.
2	EPS user state request. This is used by the HSS to request the current user status from the MME or the SGSN.
3	EPS location information request. Indicates the HSS is requesting location information from the MME or the SGSN.
4	Current location request. This bit indicates the HSS is requesting the MME or the SGSN to page the mobile device to determine its current location. This is only used with the EPS location information request bit. This is only used if the mobile device is in idle mode.
5	Locale time zone request. This bit indicates the HSS is requesting the time zone for where the mobile device is currently located.
6	Remove SMS registration. This indicates the MME should consider itself unregistered for SMS delivery.
7	RAT request. The HSS is requesting the radio access type (RAT) for the EPS location information. Only used with the EPS location request bit.
8	P-CSCF restoration request. Indicates that the HSS is requesting HSS-based P-CSCF restoration be executed.

TABLE 6.23   Bit Values for IDR-Flags AVP

IMS-Voice-Over-PS-Sessions-Supported

This is an enumerated AVP, with only two values:

   0 = Not supported

   1 = Supported

The AVP indicates whether or not the mobile devices most recent tracking area (TA) or routing area (RA) in the serving MME or SGSN support this service.

Item-Number

This is used to provide an order for requested vectors within one request for authentication vectors. It is part of the Authentication-Info AVP.

Job-Type

This enumerated AVP identifies the type of mechanism that is being used to collect measurements from mobile devices when minimization of drive tests (MDT) is being used:

   0 = Immediate MDT only

   1 = Logged MDT only

   2 = Trace only

   3 = Immediate MDT and trace

4 = Reserved
5 = RCEF reports only

KASME

This AVP is part of the EUTRAN-Authentication-Info. It provides the KASME generated by the AuC at the request of the HSS when the authentication request is received.

Kc

This AVP is part of the EUTRAN-Authentication-Info. It provides the ciphering key Kc in the format of an octet string.

Last-UE-Activity-Time

The time of the last radio contact with a mobile device as recorded by the MME or the SGSN is provided in this AVP. If the mobile device is absent, this AVP will not be present. The value is expressed as time.

LCS-Info

This AVP contains a list of gateway Mobile Location Centers (MLCs) that are authorized to send a location request for the mobile device. The privacy exception list is only applicable over the S6d interface.

<LCS-Info>::=<AVP Header: 1473, 10415>

*[GMLC-Number]
*[LCS-PrivacyException]
*[MO-LR]
*[AVP]

LCS-PrivacyException

This AVP contains the classes of LCS clients that are allowed to locate the target mobile device. This is a grouped AVP.

<LCS-PrivacyException>::=<AVP Header: 1475, 10415>

{SS-Code}
{SS-Status}
[Notification-to-UE-User]
*[External-Client]
*[PLMN-Client]
*[Service-Type]
*[AVP]

LIPA-Permission

This AVP indicates if the requested APN can be accessed through LIPA by the mobile device. LIPA allows the mobile device to use a femtocell to reach specific APNs (if allowed). This is an enumerated AVP.

0 = LIPA prohibited

1 = LIPA only
2 = LIPA is conditional (LIPA can be used, or not used)

LTE Definition							
Bits	8	7	6	5	4	3	2	1
	Spare	M5 for UL	M5 for DL	M4 for UL	M4 for DL	M3	M2	M1
Spare								
UMTS Definition							
Bits	8	7	6	5	4	3	2	1
	M7 for DL	M6 for UL	M6 for DL	M5	M4	M3	M2	M1
Spare								

TABLE 6.24   Structure for List-of-Measurements AVP

List-of-Measurements

If the Job-Type AVP value is "immediate MDT" or "immediate MDT and trace," then this AVP must be included. It defines the measurements that are to be collected. It is bit-masked, with the following definitions (Table 6.24).

LTE Values:

M1—Referenced Signal Received Power (RSRP) and Referenced Signal Received Quality (RSRQ) measurements by the mobile device with periodic, event A2 reporting triggers

M2—Power headroom (PH) measurement by mobile device (available from the MAC layer)

M3—Received interference power measurement by the eNodeB

M4—Data volume measurements separately for download and upload by the eNodeB

M5—Scheduled IP throughput measurement separately for download and upload by eNodeB

UMTS Values:

M1—CPICH RSCP and CPICH Ec/No measurement by the mobile device with periodic or event 1F reporting triggers

M2—For 1.28 Mcps TDD, P-CCPCH RSCP, and timeslot ISCP measurement by the mobile device with event 1I as reporting triggers

M3—SIR and SIR error (FDD) by eNodeB

M4—The mobile device power headroom is applicable for E-DCH transport channels

M5—Received total wideband power by NodeB in the UMTS network

M6—Data volume measurement, separately for upload and download, by the RNC in the UMTS network

M7—Upload and download throughput measurement collected by the RNC, per RAB and per mobile device

Local-Time-Zone

This is a grouped AVP providing the time zone and the adjustment required for Daylight Savings Time based on the location of the mobile device. It is provided in the IDA command when an IDR requests the information.

<Local-Time-Zone>::=<AVP Header: 1649, 10415>

{Time-Zone}

{Daylight-Saving-Time}

*[AVP]

Local-Time-Zone-Indication

This AVP is used on the Sh interface to indicate if the local time zone for the mobile device is requested with location information or without location information.

0 = Only local time zone requested

1 = Local time zone and location information requested

Location-Area-Identity

The location area identity is comprised of the mobile country code (MCC) and the mobile network code (MNC). The identifier is provided in this AVP to provide the location of the subscriber when a location request is received.

The value is formatted per ITU E.212.

Logging-Duration

Used with the MDT procedures, this AVP identifies the duration for the validity of the MDT logged configuration while the mobile device is in idle mode. It is a mandatory AVP when the job type is "logged MDT." The possible enumerated values are:

0 = 600 seconds

1 = 1200 seconds

2 = 2400 seconds

3 = 3600 seconds

4 = 5400 seconds

5 = 7200 seconds

Logging-Interval

This is an AVP used with MDT procedures. The AVP is mandatory when the job type is equal to "logged MDT." It defines the periods for which measurements are to be logged, configured in seconds and in multiples of 1.28 seconds. The AVP is enumerated with the following possible values:

0 = 1.28 seconds

1 = 2.56 seconds

2 = 5.12 seconds

3 = 10.24 seconds

8	7	6	5	4	3	2	1	
MCC digit 2				MCC digit 1				octet 1
MNC digit 3				MCC digit 3				octet 2
MNC digit 2				MNC digit 1				octet 3

TABLE 6.25   Structure of MDT-Allowed-PLMN-ID AVP

4 = 20.48 seconds

5 = 30.72 seconds

6 = 40.96 seconds

7 = 61.44 seconds

MDT-Allowed-PLMN-ID

This AVP identifies the network where measurements are to be collected as part of the MDT procedures. The identifiers are preset and provisioned in the network nodes by service providers. Table 6.25 shows the format for the network identifier, based on the MCC and MNC.

MDT-Configuration

This is a grouped AVP containing information about measurements to be recorded through minimization of drive tests (MDT). This information will be used for the configuration of this feature when it is implemented. The function allows measurements to be taken directly from mobile devices regarding network performance, in place of the method of driving around the network with test vehicles and test devices gathering measurements.

<MDT-Configuration>::=<AVP Header: 1622, 10415>

{Job-Type}

[Area-Scope]

[List-of-Measurements]

[Reporting-Trigger]

[Reporting-Interval]

[Report-Amount]

[Event-Threshold-RSRP]

[Event-Threshold-RSRQ]

[Logging-Interval]

[Logging-Duration]

[Measurement-Period-LTE]

[Measurement-Period-UMTS]

[Collection-Period-RMM-LTE]

[Collection-Period-RMM-UMTS]

[Positioning-Method]

[Measurement-Quality]

[Event-Threshold-Event-1F]

[Event-Threshold-Event-1I]

*[MDT-Allowed-PLMN-ID]

*[AVP]

MDT-User-Consent

This indicates if the subscriber has given consent for their mobile device to provide measurements when MDT is implemented in the network. The default value for this enumerated AVP is always "0."

0 = Consent not given

1 = Consent given

MIP6-Agent-Info

This AVP contains the identity of the PDN Gateway. The address format is either a domain name or an IP address.

<MIP6-Agent-Info>::=<AVP Header: 486>

*2 [MIP-Home-Agent-Address]

[MIP-Home-Agent-Host]

[MIP6-Home-Link-Prefix] (not used for S6a/S6d)

*[AVP]

Measurement-Period-LTE

This AVP is used with MDT procedures. It defines the length of time for a measurement period when the data volume and scheduled IP throughput measurements are calculated by the eNodeB. The same measurement period is used for both upload and download. This AVP is mandatory when the job type is "immediate MDT" or "immediate MDT and trace," and the List-of-Measurements AVP bits 4, 5, 6, or 7 are set to the value of 1. This is an enumerated AVP with the following possible values:

0 = 1024 ms

1 = 1280 ms

2 = 2048 ms

3 = 2560 ms

4 = 5120 ms

5 = 10,240 ms

6 = 1 minute

Measurement-Period-UMTS

This AVP is used to specify the measurement period to be used in the radio network for data volume and throughput measurements. The same value is used for both upload and download measurements.

This AVP is mandatory if the job type equals "immediate MDT" or "immediate MDT and trace," and the List-of-Measurements AVP indicates M6 or M7 for upload or download. Some values may not be supported in the network due to the large volume of measurements that would occur.

The values are all enumerated, with the following possible values:

0 = 250 ms

1 = 500 ms

2 = 1000 ms

3 = 2000 ms

4 = 3000 ms

5 = 4000 ms

6 = 6000 ms

7 = 8000 ms

8 = 12,000 ms

9 = 16,000 ms

10 = 20,000 ms

11 = 24,000 ms

12 = 28,000 ms

13 = 32,000 ms

14 = 64,000 ms

Measurement-Quantity

This bit-masked AVP describes what M1 measurements apply for event threshold 1F. It is mandatory if the job type is equal to "immediate MDT" or "immediate MDT and trace," and the trigger is set to "1F reporting." Only one of the bits can be set to a value of 1 (Table 6.26).

All other bits are reserved for future use.

MME-Location-Information

This is a grouped AVP used to provide information about the location of a mobile device as known by the MME.

<MME-Location-Information>::=<AVP Header: 1600, 10415>

[E-UTRAN-Cell-Global-Identity]

[Tracking-Area-Identity]

Bit	Description
1	CPICH Ec/No.
2	CPICH RSCP

TABLE 6.26 Bit Values for Measurement-Quantity AVP

[Geographical-Information]

[Geodetic-Information]

[Current-Location-Retrieved]

[Age-of-Location-Information]

[User-CSG-information]

*[AVP]

MME-Number-for-MT-SMS
This AVP contains the ISDN number of the MME when the MME supports SMS delivery. The ISDN number of the MME is used for routing of SMS when the MME will be used for delivery of SMS. It uses E.164 numbering format. Nature of address and numbering plan is not identified as part of this parameter.

MME-User-State
This AVP is used to provide information about the user state in the MME. It is a grouped AVP.

<MME-User-State>::=<AVP Header: 1497, 10415>

[User-State]

*[AVP]

MO-LR
This AVP contains the classes of mobile originated location requests that exist for a mobile device.

<MO-LR>::=<AVP Header: 1485, 10415>

{SS-Code}

{SS-Status}

*[AVP]

MPS-Priority
This AVP is bit-masked, with only two bits. If the value of the bit is 0, it is considered as "not set." If the value is one, the bit is considered to be set (Table 6.27).

MSISDN
The MSISDN AVP contains the MSISDN of the subscriber. It can be found in numerous commands where the subscriber's public identity is used. The MSISDN is represented in this AVP as an octet string.

Bit	Definition
0	MPS circuit-switched priority. Indicates the mobile device is subscribed to an eMLPP or 1xRTT priority service in the circuit-switched network.
1	MPS EPS priority. Indicates that the mobile device is subscribed in the evolved packet system (EPS)

TABLE 6.27 Bit Values for MPS-Priority AVP

Network-Access-Mode

This AVP is used by the HSS to identify what type of access is allowed for a specific subscription. The enumerated values are:

0 = Packet and circuit switched access

1 = Reserved

2 = Only packet

NOR-Flags

This AVP is bit-masked. Each of these bits can be set to a value of 0 or 1. The below values assume the bit is set to the value of 1 (Table 6.28).

Notification-to-UE-User

This is an enumerated AVP, with the following values:

0 = Notify location allowed

1 = Notify and verify location allowed if no response

2 = Notify and verify not allowed if no response

3 = Location not allowed

Bit	Description
0	Single registration indication. This indicates that the HSS will send a Cancel-Location to the SGSN. It is set only by the HSS and included in the NOR command.
1	SGSN are restricted. This bit indicates the complete SGSN area is restricted because of a regional subscription.
2	Ready for SM from SGSN. This bit indicates the mobile device is present and able to receive SMS traffic from the SGSN.
3	Device reachable from the MME. This bit indicates the mobile device has become reachable by the MME.
4	Reserved, and not used.
5	Mobile device reachable from the SGSN. This bit indicates the mobile device has become unreachable by the SGSN.
6	Ready for SM from MME. This bit indicates the mobile device is present and able to receive SMS traffic from the MME.
7	Homogenous support of IMS voice over PS sessions. This bit indicates that this has been updated.
8	S6a/S6d indicator. This bit indicates the NOR was sent on the S6a interface by the MME if set to the value of 1. If set to the value of 0, indicates the NOR was sent on the S6d interface by the SGSN.
9	Removal of MME registration for SMS. This bit indicates the MME requests to remove its registration for SMS if the value is set to 1.

TABLE 6.28 Bit Values for NOR-Flags AVP

Number-of-Requested-Vectors

This AVP is found in the Requested-UTRAN-GERAN-Authentication-Info and Requested-EUTRAN-Authentication-Info grouped AVPs. It identifies the number of authentication vectors that are being requested by an MME or SGSN.

OMC-ID

This is an optional AVP used to identify an operation and maintenance center in the operator network. The value is an octet string.

One-Time-Notification

When present, this AVP indicates the sender is requesting a one-time notification for mobile device reachability. This is used on the Sh interface.

0 = One time notification requested

Operator-Determined-Barring

Each bit identifies a specific service to be barred. When set to the value of 1, indicates service is barred from the subscriber (Table 6.29).

PDN-GW-Allocation-Type

This AVP indicates if the PDN Gateway address found in the MIP6-Agent-Info AVP has been assigned dynamically or statically. There are only two enumerated values:

0 = Static

1 = Dynamic

PDN-Type

This AVP indicates the type of address for the packet data network. Only the indicated address type can be used to access the packet data network. This is only used as part of the APN subscription context. The possible enumerated values are

Bit	Description
0	All packet services are barred
1	Roamer access to the home PLMN AP is barred
2	Roamer access to the visited PLMN AP is barred
3	Barring of all outgoing calls
4	All outgoing international calls are barred
5	All outgoing international calls except those to the home country are barred
6	All outgoing interzonal calls are barred
7	All outgoing interzonal calls except those to the home country are barred
8	All outgoing international calls and interzonal calls except those to the home country are barred

TABLE 6.29   Bit Values for Operator-Determined-Barring AVP

0 = IPv4

1 = IPv6

2 = IPv4 and IPv6

3 = IPv4 or IPv6

PDP-Context

The subscriber can have multiple GPRS PDP contexts. Even if they subscriber does have multiple PDP contexts, the Service-Selection AVP can be the same for each different PDP context.

<PDP-Context>::=<AVP Header: 1469, 10415>

{Context-Identifier}

{PDP-Type}

[PDP-Address]

{QoS-Subscribed}

[VPLMN-Dynamic-Address-Allowed]

{Service-Selection}

[3GPP-Charging-Characteristics]

[Ext-PDP=Type]

[AMBR]

[APN-OI-Replacement]

[SIPTO-Permission]

[LIPA-Permission]

[Restoration-Priority]

[SIPTO-Local-Network-Permission]

*[AVP]

PDP-Type

This AVP identifies the version of IP supported by the mobile device: IPv4, IPv6, or a dual stack of IPv4v6. If both IPv4 and IPv6 are supported, but not through dual stack support, this must be sent twice for the same APN. One instance indicates support of IPv4 and the other indicates support of IPv6. The values are expressed in hexadecimal values. IPv4 is expressed as hexadecimal 21, while IPv6 is expressed as hexadecimal 57.

PLMN-Client

This AVP is enumerated with the following values:

0 = Broadcast service

1 = O and M HPLMN

2 = O and M VPLMN

3 = Anonymous location

4 = Target mobile device subscribed service

Bit	Description
1	GNSS. This can only be valid if M1 measurement is configured in the List-of-Measurements AVP.
2	E-Cell ID.

TABLE 6.30 Bit Values for Positioning-Method AVP

Positioning-Method

This AVP is used with the MDT procedures. It defines the positioning method that should be used, but only applies to LTE networks. This is used when the job type is equal to "immediate MDT" or "immediate MDT and trace." It is a bit-masked AVP with two bits defined (Table 6.30).

All other bits (3 to 8) in this AVP are reserved. If both bits are set to the value of 1, the eNodeB shall use E-Cell ID measurement collection only if the mobile device does not support GNSS-based location information.

PUA-Flags

This is a bit-masked AVP, used to indicate how the M-TMSI or P-TMSI are to be handled after a PUR (Table 6.31).

Public-Identity

This AVP is used on the Sh interface to identify the SIP or TEL URL of an IMS user.

PUR-Flags

This is a bit-masked AVP with two bits. The values given here assume the value of the bit is set to one. The bits indicate if the mobile device profile has been purged in either the MME or the SGSN, but is only used when the MME and the SGSN are a combined node (Table 6.32).

Bit	Description
0	Freeze the M-TMSI. This indicates the M-TMSI value should be frozen in the MME, and not made available for immediate reuse.
1	Freeze the P-TMSI. This indicates the P-TMSI value should be frozen in the SGSN, and not made available for immediate reuse.

TABLE 6.31 Bit Values for PUA-Flags AVP

Bit	Description
0	Purged in the MME
1	Purged in the SGSN

TABLE 6.32 Bit Values for PUR-Flags AVP

Pre-Paging-Supported

This AVP is used on the Sh interface to indicate if the sender supports pre-paging or not. There are two enumerated values:

0 = Pre-paging not supported

1 = Pre-paging supported

Prose-Subscription-Data

Contains data from the proximity services subscription data. This is a grouped AVP.

<Prose-Subscription-Data>::=<AVP Header: xxx, 10415>

{Prose-Permission}

*[AVP]

QoS-Subscribed

This is used in the PDP-Context grouped AVP, providing the QoS supported in the subscription. The value is provided in an octet string.

RAND

The RAND is part of the authentication vectors provided by the HSS when the MME or the SGSN request authentication information. It is expressed as an octet string.

RAT-Frequency-Selection-Priority-ID

The network uses the RAT frequency selection priority (RFSP) index to identify radio channels to be used as part of the networks radio resource management. The MME receives this index from the HSS for a subscriber, and sends this information to the eNodeB (using the non-Diameter S1 interface).

This AVP is used by the HSS to send the information to the MME during an attach procedure, where it is stored as part of the subscription data. The value is a number from 1 to 256.

Regional-Subscription-Zone-Code

A zone code represents a collection of tracking or routing areas where the subscriber is allowed to roam. The MME and/or the SGSN are responsible for determining if the subscriber is allowed to roam or not in a specific zone, based on their subscription data. Up to 10 zone codes can be identified in subscription data. This information is sent in the Subscription-Data AVP by the HSS.

Relay-Node-Indicator

This AVP indicates whether or not the subscription data belongs to a relay node. It is an enumerated AVP with two possible enumerated values:

0 = Not a relay node

1 = A relay node

Report-Amount

This is used with MDT procedures. It defines how many measurement reports will be taken while the mobile device is connected. The AVP is mandatory when the reporting trigger is set for periodic measurements from the mobile device and the job type is "immediate MDT" or "immediate MDT and trace." This is an enumerated AVP with the following possible values:

0 = 1

1 = 2

2 = 4

3 = 8

4 = 16

5 = 32

6 = 64

7 = infinity

Report-Interval

This AVP is mandatory when the Report-Trigger is set for periodic mobile device measurements and the Job-Type AVP has the value of "immediate MDT" or "immediate MDT and trace." It indicates the intervals between reports for periodic measurements whenever the mobile device is connected. The first set of values represents UMTS values, while the last half are for LTE values. This is an enumerated AVP.

UMTS Values:

0 = 250 ms

1 = 500 ms

2 = 1000 ms

3 = 2000 ms

4 = 3000 ms

5 = 4000 ms

6 = 6000 ms

7 = 8000 ms

8 = 12,000 ms

9 = 16,000 ms

10 = 20,000 ms

11 = 24,000 ms

12 = 28,000 ms

13 = 32,000 ms

14 = 64,000 ms

LTE Values:

15 = 120 ms

16 = 240 ms

17 = 480 ms

18 = 640 ms

19 = 1,024 ms

20 = 2,048 ms

21 = 5,120 ms

22 = 10,240 ms

23 = 60,000 ms (1 minute)

24 = 360,000 ms (6 minutes)

25 = 720,000 ms (12 minutes)

26 = 1,800,000 ms (30 minutes)

27 = 3,600,000 ms (60 minutes)

Reporting-Trigger

This AVP is mandatory when the Job-Type AVP contains the value of "immediate MDT" or "immediate MDT and trace." It is also required when measurements are from the mobile device. You cannot have reporting period combined as periodic event-based and periodic, event-based, or event-based periodic reporting only (Table 6.33).

Repository-Data-ID

This AVP is used on the Sh interface to identify the Service-Indication and Sequence-Number AVPs for requested repository data.

<Repository-Data-ID>::=<AVP Header: 715, 10415>

{Service-Indication}

{Sequence-Number}

*[AVP]

Bit	Value when Set
1	Periodical
2	Event A2 for LTE
3	Event 1F for UMTS
4	Event 1I for UMTS 1.28Mcps TDD
5	A2 event triggered periodic for LTE
6	All configured RRM event triggers for LTE
7	All configured RRM event triggers for UMTS
8	Reserved

TABLE 6.33   Bit Values for Reporting-Trigger AVP

Reset-ID

This optional AVP is used to identify a resource within the HSS that may have failed or restarted. The MME or the SGSN use this along with the HSS realm to determine which subscribers are impacted by the failure, and mark the subscription records accordingly.

Restoration-Priority

The Restoration-Priority indicates the priority to be used when restoring a subscribers data connections caused by the failure or restart of the SGSN or PDN Gateway. The value can be a number between 1 and 16, with 1 representing the highest priority.

Result-Code

The Result-Code AVP is used to either indicate success of a command execution, or communicate an error. It is also used in conjunction with the Experimental-Result AVP, which provides codes specific to a vendor (such as 3GPP). The result codes used in this AVP can be found in Chap. 3. For the sake of this application, the result codes are the same as described in that chapter. For result codes specific to 3GPP on S6a/S6d, look at the definition in this chapter for the Experimental-Result AVP.

Re-Synchronization-Info

The HSS sends this AVP with an octet string derived from a concatenated RAND and AUTS.

Requested-Domain

This AVP is used on the Sh interface to indicate which domain data is being requested for, circuit-switched or packet-switched. The values are enumerated:

0 = Circuit-switched domain

1 = Packet-switched domain

Requested-EUTRAN-Authentication-Info

This AVP is a grouped AVP used by the MME to request authentication information from the HSS.

< Requested-EUTRAN-Authentication-Info>::=<AVP Header: 1408, 10415>

[Number-of-Requested-Vectors]

[Immediate-Response-Preferred]

[Re-Synchronization-Info]

*[AVP]

Requested-UTRAN-GERAN-Authentication-Info

This AVP is a grouped AVP used by the SGSN to request authentication credentials from the HSS.

< Requested-UTRAN-GERAN-Authentication-Info>::=<AVP Header: 1409, 10415>

[Number-of-Requested-Vectors]

[Immediate-Response-Preferred]

[Re-Synchronization-Info]

*[AVP]

Roaming-Restricted-due-to-Unsupported-Feature

This is an optional AVP indicating that roaming is not allowed because of an unsupported feature. There is only one enumerated value:

0 = Roaming restricted

Routing-Area-Identity

The Routing-Area-Identity is based on the routing area code, the local area code, and MNC/MCC. It uses the form of a subdomain. By adding the prefix "rac" to the front of the home network realm. It might look something like this:

Rac.lac.rac.epc.mnc.mcc.3gppnetwork.org

Send-Data-Indication

This AVP indicates that the sender is requesting user data. The values are enumerated:

0 = User data is not being requested

1 = User data is being requested

Sequence-Number

This AVP is used on the Sh interface when repository data is being sent to represent the sequence number for the repository data.

Server-Name

This AVP contains the SIP URL identifying the application server in the IMS network.

Service-Area-Identity

Cells are sometimes grouped in a service area, depending on the service provider. Each service area is then given a unique identifier. This AVP is used to send the service area ID currently serving the subscriber.

Service-Indication

This AVP is used on Sh to identify a service or a set of services supported in an application server. The services are presented as an octet string.

Service-Type

This AVP contains the service type of the clients allowed to locate a target mobile device for a mobile terminated location request.

<Service-Type>::=<AVP Header: 1483, 10415>

{Service-TypeIdentity}

[GMLC-Restriction]

[Notification-to-UE-User]

*[AVP]

Service-TypeIdentity

The service type is used to identify the type of service that may be requesting location information for a subscriber. The Service-Type and the Service-TypeIdentity are used together. Following are the values for this enumerated AVP:

0 = Emergency services

1 = Emergency alert services

2 = Person tracking

3 = Fleet management

4 = Asset management

5 = Traffic congestion reporting

6 = Roadside assistance

7 = Routing to nearest commercial enterprise

8 = Navigation

Serving-Node-Indication

This AVP is used on the Sh interface to indicate that only the serving nodes need to be identified in the location information being requested. No other location information is required. The value is enumerated:

0 = Only serving nodes required

Session-Priority

The session priority is used to identify high-priority sessions to the call session control function (CSCF) and the HSS. The values are operator defined, and are enumerated in this AVP. This is used on the Sh interface.

0 = Priority 0

1 = Priority 1

2 = Priority 2

3 = Priority 3

4 = Priority 4

SGSN-Location-Information

This is a grouped AVP containing information about the location for a mobile device based on the SGSN knowledge.

< SGSN-Location-Information>::=<AVP Header: 1601, 10415>

[Cell-Global-Identity]

[Location-Area-Identity]

[Service-Area-Identity]

[Routing-Area-Identity]

[Geographical-Information]

[Geodetic-Information]

[Current-Location-Retrieved]

[Age-of-Location-Information]

[User-CSG-Information]

*[AVP]

SGSN-Number

This AVP provides the ISDN number for the SGSN requesting location information. The number is formatted in E.164 format.

SGSN-User-State

This AVP contains information about the user state in the SGSN. It is a grouped AVP.

<SGSN-User-State>::=<AVP Header: 1498, 10415>

[User-State]

*[AVP]

SIPTO-Local-Network-Permission

Used to indicate if traffic associated with this APN is allowed to use Selected IP traffic offload (SIPTO) in the local network. It is enumerated with the following potential values:

0 = SIPTO in the local network allowed

1 = SIPTO in the local network not allowed

SIPTO-Permission

This AVP will indicate if SIPTO can be used for the specified APN. It is an enumerated AVP with the following values:

0 = SIPTO above RAN is allowed

1 = SIPTO above RAN is not allowed

SMS-Register-Request

This AVP indicates if the MME or the SGSN require they be registered for SMS services, or if they prefer not to be registered. The AVP is enumerated with the following possible values:

0 = SMS registration is required

1 = SMS registration not preferred

2 = No preference

Software-Version

The Software-Version is a two-digit number determined by the equipment vendor. It reflects the software version currently running on the mobile device. The software version is part of the IMEI for a device, which is comprised of the type allocation code (TAC), serial number, and the software version.

Specific-APN-Info

This AVP is used in the APN configuration when an APN wild card is used. It contains the APN the mobile device is authorized to connect to the registered PDN Gateway.

<Specific-APN-Info>::=<AVP Header: 1472, 10415>

{Service-Selection}

{MIP6-Agent-Info}

[Visited-Network-Identifier]

*[AVP]

SRES

SRES is part of a triplet used in authentication. A triplet consists of the network challenge (RAND), the user response (SRES), and a cipher key (Kc). The value in this AVP is in the form of an octet string.

SS-Code

This AVP identifies the supplementary services to be deleted from a subscription. The value is in the form of an octet string.

SS-Status

The SS-Status AVP gives the status of supplementary services. It is an octet string representing the following (Table 6.34):

Bits 8765: 0000 (unused)

Bit 4: Q bit

Bit 3: P bit

Bit 2: R bit

Bit 1: A bit

STN-SR

The session transfer number is used for SRVCC, and is included in the subscriber data. This AVP is used to indicate the session transfer number is to be deleted from the subscriber data. The value is an international number following E.164 formatting.

	P Bit	R Bit	A Bit	Q Bit
Provisioned	1			
Not provisioned	0			
Registered		1		
Not registered		0		
Active and operative			1	0
Active and quiescent			1	1
Not active			0	0/1

TABLE 6.34 Bit Values for SS-Status AVP

Bit	Description
0	Subscribed to SRVCC

TABLE 6.35 Bit Values for Subscribed-VSRVCC AVP

Subscribed-VSRVCC

Indicates if the subscriber has subscribed to vSRVCC services. If this AVP is not present in the ULA command, it means the subscriber is not subscribed to SRVCC. It is a bit-masked AVP with one bit. If the bit is set to the value of 1, it has the following meaning (Table 6.35):

Subscriber-Status

This AVP is used to convey if service is to be granted or if barring has been implemented. The values are enumerated.

0 = Service granted

1 = Operator Determined Barring (ODB)

Subscribed-Periodic-RAU-TAU-Timer

The routing area update and tracking area update timers are stored in the mobile device, and communicated via location updates. This AVP contains the value of the RAU/TAU timer in seconds.

Subscription-Data

This is a grouped AVP containing the users profile for evolved packet services and GERAN/UTRAN. The APN-Configuration-Profile AVP will contain the profile for the default APN configured for this subscriber, when applicable.

The AMBR AVP must also be included when the APN-Configuration-Profile AVP is present in the ULA. This AVP is also used when the EPS-Subscribed-QoS-Profile AVP is used in the APN-Configuration AVP.

If the Subscriber-Status AVP is present with the value of "OPERATOR_ DETERMINED_BARRING," then the Operator-Determined-Barring AVP or the HPLMN-ODB AVP must also be included.

The Access-Restriction-Data AVP is only included if any of the restrictions apply to this subscriber. Otherwise the AVP is omitted.

This AVP can be skipped if the ULR-Flags in the ULR indicate subscriber data can be skipped.

<Subscription-Data>::=<AVP Header: 1400, 10415>

[Subscriber-Status]

[MSISDN]

[A-MSISDN]

[STN-SR]

[ICS-Indicator]

[Network-Access-Mode]

[Operator-Determined-Barring]

[HPLMN-ODB]

*10 [Regional-Subscription-Zone-Code]

[Access-Restriction-Data]

[APN-OI-Replacement]

[LCS-Info]

[Teleservice-List]

*[Call-Barring-Info]

[3GPP-Charging-Characteristics]

[AMBR]

[APN-Configuration-Profile]

[RAT-Frequency-Selection-Priority-ID]

[Trace-Data]

[GPRS-Subscription-Data]

[CSG-Subscription-Data]

[Roaming-Restricted-due-to-Unsupported-Feature]

[Subscribed-Periodic-RAU-TAU-Timer]

[MPS-Priority]

[VPLMN-LIPA-Allowed]

[Relay-Node-Indicator]

[MDT-User-Consent]

[Subscribed-VSRVCC]

[ProSe-Subscription-Data]

[Subscription-Data-Flags]

*[AVP]

Subscription-Data-Flags

This AVP identifies how the subscription data is to be applied. It is bit-masked, with two bits defined. Each bit represents the following (Table 6.36):

A value of 1 in the first bit indicates the subscription is only for packet services. Circuit-switched services are only permitted for the delivery of SMS messages.

A value of 1 in the second bit indicates the subscription permits SMS to be delivered via the SGSN, so circuit-switched services are not needed for SMS delivery.

Bit	Description
0	Packet services and SMS only
1	SMS in SGSN allowed

TABLE 6.36 Bit Values for Subscription-Data-Flags AVP

Subs-Req-Type

The Subs-Req-Type indicates what type of notifications are being subscribed to by the application server in the IMS network. This AVP is used on the Sh interface. The values are enumerated:

0 = Subscribe to notification of changes in subscriber data

1 = Unsubscribe to notifications of changes in subscriber data

Supported-Applications

This grouped AVP identifies the applications that are supported by a Diameter node.

<Supported-Applications>::=<AVP Header: 631, 10415>

*[Auth-Application-ID]

*[Acct-Application-ID]

*[Vendor-Specific-Application-ID]

*[AVP]

Supported-Features

This is a grouped AVP that provides a list of supported features. The Vendor-ID and the Feature-List-ID AVPs are used collectively to specify the list of features based on the vendor of the network element. The 3GPP vendor ID is always 10415.

<Supported-Features>::=<AVP Header: 628, 10415>

{Vendor-ID}

{Feature-List-ID}

{Feature-List}

*[AVP]

Teleservice-List

Telesevice-List contains the service codes for short message services for a subscriber.

<Teleservice-List>::=<AVP Header: 1486, 10415>

1*{TS-Code}

*[AVP]

Terminal-Information

This is a grouped AVP used to provide information about the subscriber's device. Only the IMEI and the software version are included on the S6a/S6d/S13/S13' interface.

<Terminal-Information>::=<AVP Header>: 1401, 10415>

[IMEI]

[3GPP-MEID]

[Software-Version]

*[AVP]

Time-Zone

This AVP contains the local time zone where the mobile device is connected. The time zone is represented as an offset from UTC (coordinated universal time) expressed in units of 15 minutes. A character string (such as +8) is used to represent the time zone.

Trace-Collection-Entity

The trace collection entity is where the mobile device sends trace data. The address for the trace collection entity is an IP address, either IPv4 or IPv6.

> Trace-Data
>
> <Trace-Data>::=<AVP Header: 1458, 10415>
>
> {Trace-Reference}
>
> {Trace-Depth}
>
> {Trace-NE-Type-List}
>
> {Trace-Interface-List}
>
> {Trace-Event-List}
>
> [OMC-ID]
>
> {Trace-Collection-Entity}
>
> [MDT-Configuration]
>
> *[AVP]

Trace-Depth

This AVP is enumerated, and defines whether or not entire signaling messages or just some elements are to be recorded as part of MDT procedures. There are several possible values:

0 = Minimum—record some of the signaling elements as well as any vendor specific extensions

1 = Medium—record some of the signaling elements and radio measurement elements, plus any vendor specific extensions

2 = Maximum—record all signaling message elements plus any vendor specific extensions

3 = Minimum without vendor specific extensions—same as 0 without the extensions

4 = Medium without vendor specific extensions—same as 1 without the extensions

5 = Maximum without vendor specific extensions—same as 2 without the extensions

Trace-Event-List

This is part of the MDT process, and is expressed as an octet string.

Trace-Interface-List

This is part of the MDT process, and is expressed as an octet string.

Trace-NE-Type-List

This is part of the MDT process, and is expressed as an octet string.

Trace-Reference
This is part of the MDT process, and is expressed as an octet string. It contains a concatenation of the MCC, MNC, and trace ID.

Tracking-Area-Identity
The tracking area ID identifies the tracking area where the subscriber is currently located and connected. It consists of the MCC, MNC, and TAC, expressed in the form of a fully qualified domain name (FQDN). When expressed as FQDN, the format looks like:

<TAC low byte> . <TAC high byte> . <MNC> . <MNN> . 3gppnetwork.org

Ts-Code
Used in the Teleservice-List AVP, this AVP identifies the teleservice codes to be deleted from a subscription.

UDR-Flags
The UDR flags are used on the Sh interface. This is a bit-masked AVP, so when the bit is set to the value of 1, it will have the following meaning (Table 6.37):

UE-SRVCC-Capability
This is an enumerated AVP indicating whether or not the mobile device supports single radio voice call continuity (SRVCC). There are only two values:

0 = The mobile device does not support SRVCC.

1 = The mobile device does support SRVCC.

ULA-Flags
This AVP is bit-masked, with the following values (Table 6.38):
All other bits are ignored.

Bit	Description
0	This is used to indicate that the EPS location information can be sent to the application server when circuit-switched location information is requested.
1	Indicates the radio access type is being requested when the packet-services location information is requested.

TABLE 6.37   Bit Values for UDR-Flags AVP

Bit	Description
0	Separation indication, indicating the HSS stores the SGSN number and the MME number in separate memory
1	MME registered for SMS, indicating the HSS has registered the MME for SMS

TABLE 6.38   Bit Values for ULA-Flags AVP

Bit	Description
0	Single registration indication. When this bit is set to the value of 1, the HSS will send a Cancel-Location to the SGSN.
1	S6a/S6d indicator. When the value is set to 1, it means the ULR was sent on the S6a interface by the MME. When the value is set to 0, it indicates the ULR was sent by the SGSN on the S6d interface.
2	Skip subscriber data. When the value is set to 1, it indicates the Subscription-Data AVP can be omitted from the ULA.
3	GPRS subscription data indicator. If this value is 1, the GPRS subscription data must be provided in the ULA (if it is known). If the value is 0, GPRS subscription data is only included if the ULR was sent by the SGSN (on the S6d) and there is no APN configuration for the subscriber.
4	Node type indicator. Only set to the value of 1 if the ULR was sent by a combined MME/SGSN node.
5	Initial attach indicator. Indicates that the HSS will send a Cancel-Location to the SGSN or the MME.

TABLE 6.39   Bit Values for ULR-Flags AVP

ULR-Flags

This AVP is bit-masked, and each bit is defined in Table 6.39.

All other bits in this AVP are sent with a value of 0 and are discarded by the receiving HSS.

User-ID

This AVP contains the subscriber ID, containing the leading digits of the IMSI. The HSS uses this to identify a set of subscribers with the same leading IMSI digits.

User-Identity

This AVP is different from the previous one. In this AVP, the subscriber MSISDN or their public identity is provided as a grouped AVP.

<User-Identity>::=<AVP Header: 700, 10415>

[Public-Identity]

[MSISDN]

*[AVP]

User-Name

This AVP carries the IMSI of the subscriber. The IMSI is formatted to identify the home country of the subscriber, their home network, and their individual subscriber ID. It is not more than 15 digits in length in the following format (Table 6.40):

User-Data

This octet string represents the user data being requested in the UDR, SNR, and PNR commands. The user data in this AVP is to be modified when sent in the PUR command. User data sent in this AVP is used within the IMS domain for conveying the state of the subscriber, location, and other subscription information.

Mobile Country Code 3 digits	Mobile Network Code 2 to 3 digits	Mobile Subscriber Identification Code 9 to 10 digits

TABLE 6.40 Structure for User-Name AVP

User-State
This AVP is enumerated, and is used to indicate the user state in the MME or in the SGSN depending on which AVP it is used in. For example, if it is used within the MME-User-State AVP, it will provide information from the MME regarding the user's state. The following values are supported:

0 = Detached. The mobile device is in EMM_Deregistered state.

1 = Attached but not reachable for paging. The SGSN has determined that the mobile device is attached, but it cannot be reached through paging. This is only applicable to S4-SGSNs.

2 = Attached and reachable for paging. The SGSN has determined that the mobile device is attached, but no bearer channel is active. The mobile device is reachable through paging. Applies to S4-SGSNs.

3 = Connected but not reachable for paging. The SGSN or the MME have determined that the mobile device is attached and there is at least one active bearer channel assigned, but the mobile device cannot be paged.

4 = Connected and reachable for paging. The SGSN or the MME has determined that the mobile device is attached to the network and there is at least one bearer channel active. The mobile device can therefore be reached via paging.

5 = Reserved.

UTRAN-Vector
This AVP is used by the HSS to send authentication credentials to the requesting SGSN. This is a grouped AVP.

< UTRAN-Vector>::=<AVP Header: 1415, 10415>

[Item-Number]

{RAND}

{XRES}

{AUTN}

{Confidentiality-Key}

{Integrity-Key}

*[AVP]

UVA-Flags
This AVP is bit-masked with the following values (Table 6.41):
All other bits are ignored.

UVR-Flags
This AVP is bit-masked with only one value (Table 6.42):

Bit	Description
0	Temporary empty VPLMN CSG subscription data—indicating the CSS does not currently have subscription data for the subscriber, but has registered the MME and SGSN in the event a subscription occurs at a later time

TABLE **6.41**   Bit Values for UVA-Flags AVP

Bit	Description
0	Skip subscriber data—indicating the CSS may skip sending subscriber data in the UVA command.

TABLE **6.42**   Bit Values for UVR-Flags AVP

Visited-PLMN-ID

This AVP contains the visited PLMN ID when a subscriber is roaming. It is comprised of the MCC and MNC in the form of an octet string.

VPLMN-CSG-Subscription-Data

This is a grouped AVP, providing the CSG subscription ID and optionally an expiration date and time.

> VPLMN-CSG-Subscription-Data ::= <AVP header: 1641 10415>
>
> {CSG-Id}
>
> [Expiration-Date]
>
> *[AVP]

VPLMN-Dynamic-Address-Allowed

This AVP is found in the APN-Configuration AVP and indicates if the mobile device is allowed to use the PDN Gateway in its home network or visited network. If this AVP is not present, the default is the mobile device can only use the PDN Gateway in its home network. There are only two enumerated values:

> 0 = Not allowed
>
> 1 = Allowed

VPLMN-LIPA-Allowed

Local IP access (LIPA) allows mobile devices to connect to femtocells or to the packet network, based on rules and policy in the network. This AVP pertains to the permission to support LIPA when the mobile device is in the visited network. It is an enumerated AVP with only two values:

> 0 = LIPA is not allowed
>
> 1 = LIPA is allowed

Wildcarded-IMPU
The wildcarded IP Multimedia public user identity is in the form of a SIP or TEL URL. This AVP is used on the Sh interface.

Wildcarded-Public-Identity
There are some instances where a group of public identities are treated in the same fashion by the network, because they share the same service profiles, and may be included in the same registration set. Storing them as wildcarded public identities is a means of optimizing the performance of the HSS. The format is in the form of a SIP URL. This AVP is used on the Sh interface.

WLAN-Offloadability
This grouped AVP contains information regarding the ability to offload the traffic from the specific APN off of the LTE network and onto another network.

<WLAN-Offloadability>::=<AVP Header: 1667>

[WLAN-Offloadability-EUTRAN]

[WLAN-Offloadability-UTRAN]

*[AVP]

WLAN-Offloadability-EUTRAN
This AVP is specific to EUTRAN networks. It indicates whether or not the traffic can be offloaded from the network. It is bit-masked as follows (Table 6.43):

WLAN-Offloadability-UTRAN
This AVP is specific to UTRAN networks. It indicates whether or not the traffic can be offloaded from the network. It is bit-masked as follows (Table 6.44):

XRES
The XRES is part of authentication. This AVP provides the XRES during the authentication process.

Bit	Description
0	Indicates traffic with the specified APN can be offloaded to the WLAN if the bit is set to the value of 1. If the bit is cleared (value is 0) than the traffic cannot be offloaded.

TABLE **6.43**　Bit Values for WLAN-Offloadability-EUTRAN AVP

Bit	Description
0	Indicates traffic with the specified APN can be offloaded to the WLAN if the bit is set to the value of 1. If the bit is cleared (value is 0), the traffic cannot be offloaded.

TABLE **6.44**　Bit Values for WLAN-Offloadability-UTRAN AVP

CHAPTER 7

Connecting to Policy

W hen a mobile device establishes a data connection in the wireless network, the user traffic is routed from the Mobility Management Entity (MME) to the packet core. The Serving GPRS Support Node (SGSN) is the first node to receive the data packets and routes these packets either internally or to a gateway to connect with the Internet or other destinations. There is limited control over how these data connections are handled in 3G networks, which is why the 3GPP defined a new function for the packet core.

The Policy and Charging Rules Function (PCRF) is a function defined for LTE networks that allow service providers to control data sessions based on a number of triggers. These triggers can be based on the type of session, the type of device, and many other factors that we will discuss in more detail throughout this chapter.

Rules are defined in network nodes (such as the SGSN or PDN Gateway) by the PCRF when specific events occur (triggers). The network node can notify the PCRF of a data session, requesting for rules to be provisioned, or the PCRF can activate rules already provisioned in the node. We will discuss the specifics about how this is done as we talk about each of the interfaces.

There are two methods or modes used by PCRF for provisioning rules. Rules can be pushed or pulled. When the PCRF is sending rules to a network element without a request, the rule is being pushed. When the network element is requesting a rule, it will pull the rule from the PCRF. We will be using this terminology throughout this chapter.

The interface used is based on the network element that is attempting to communicate with the PCRF. Each defined interface defines the procedures used for provisioning rules in the network nodes and even determines the commands and AVPs used. We will describe several of the most commonly used interfaces in LTE networks, but there are several we have not covered in this edition of the book.

Gates can be defined within a rule to instruct the node if traffic is allowed to flow or is to be discarded. Gates are defined as either open or closed.

The implementation of the PCRF has enabled service providers to control subscriber's data sessions based on a number of factors. However, PCRF is not limited to 4G networks. Many service providers have begun implementing PCRF in their 3G networks as well, using Diameter in 3G domains. The 3GPP has defined a number of interfaces for the use of policy management and Diameter in 3G networks.

Network nodes use filters to determine what triggers a rule to be implemented (or provisioned) and templates for groups of rules based on the type of sessions and other characteristics. When a device is connected, it establishes an IP-CAN session. There may be multiple data flows within one IP-CAN session, but all associated with the same subscriber. Filters can be applied to data flows or IP-CAN sessions.

A data flow in an IP-CAN session identifies specific connections assigned to a subscriber or mobile device. It is often defined as the path that data takes from the origin to the destination, and consists of packets of user data. Think of a data flow as the equivalent to a telephone call connection, but in the data world. The data flow is from the device to the specific application being used by the subscriber (such as YouTube or Facebook).

The IP data flow is used for the purpose of routing and provisioning rules in the case of policy. Each data flow is given an identifier, which is used by the PCRF when provisioning rules for a data flow. Rules are not always assigned to a data flow. In some cases rules may be assigned to traffic sent to a specific application, which is managed by the traffic detection function (TDF) explained later in this chapter. So rules can be assigned to an entire IP-CAN session (which would be applied to all the data flows within that session), to individual data flows, or to applications, or all of the above.

The direction of traffic in a data flow uses the reference uplink and downlink. Think of uplink as the sending or output from of a network node, and downlink as receiving or input. We will refer to uplink and downlink throughout this chapter.

The network node uses filters to determine if a rule needs to be applied to a specific data flow. Templates identify the rules to be applied to specific IP-CAN bearer channels and to set the criteria for when rules apply.

Event triggers are used to identify when the network node should request new rules or modification to existing rules from the PCRF. They are defined using the Event-Trigger AVP and sent by the PCRF to the network node using the command defined for the associated interface. If a session is a data session and the traffic is connecting to an enforcement point (SGSN or PDN Gateway) then the enforcement point will send a request using the Credit Check Request (CCR) command and the PCRF will answer the request using the Check Credit Answer (CCA) command.

If the data session is a voice session connecting to the IP Multimedia Subsystem (IMS), the call session control function (CSCF) in the IMS will notify the PCRF and request rules for the session. The PCRF will then forward rules to the packet core as well, since the voice is being sent through the packet network and requires a packet connection at the enforcement point. The PCRF will then use the Re-Auth-Request/ Answer (RAR) command to the enforcement point for the session. More on this as we discuss each interface.

Rules can be dynamic or predefined. A predefined rule is provisioned in the network enforcement point. The rule can be changed, activated, or deactivated by the PCRF at any time. Dynamic rules are created in the PCRF and sent to the enforcement point.

This chapter deals with connecting to the PCRF for the purpose of provisioning of the enforcement points and interfacing with the charging platforms for billing.

The Policy and Charging Rules Function

The PCRF does not enforce policies. The actual subscriber traffic (bearer traffic) does not go through the PCRF. Enforcement of policy is left to the network elements that manage the bearer traffic. This includes functions such as the MME, the SGSN, the TDF, CSCF, and the PDN Gateway to name a few. The PCRF is simply where rules are defined for the subscriber. It provides a central place for the creation and management of all rules in the network.

The service provider defines policy profiles that are stored in the subscriber profile repository (SPR), or the user data repository (UDR). The SPR has evolved to what we now refer to as the UDR/SPR, which is based on the User Data Convergence model defined by 3GPP. The PCRF uses this repository as a database for policy profiles. The PCRF accesses the UDR/SPR via the Ud interface, while the SPR is accessed using the Sp interface.

The enforcement points query the PCRF when a subscriber or mobile device attempts to make a connection to the network (for either voice or data). Each enforcement function will query the PCRF for its part of the session. For example, when user traffic is sent to the SGSN, the SGSN will query the PCRF for rules specific to how it should handle the traffic. The SGSN may then route the traffic to a TDF, which will also query the PCRF and request rules specific to how it should manage the traffic. Since each network element provides different capabilities and functions, the rules and the information shared between the PCRF and the enforcement point will be different and based on the interface.

The PCRF accesses the HSS to retrieve subscription information and it interacts with charging (especially prepaid charging) to report usage and to ensure the subscriber does not use services for which they have not paid. Policy and charging are critical to managing services in a prepaid network, ensuring the subscriber uses the services and bandwidth they are entitled to according to their subscription. We will discuss these procedures later.

Figure 7.1 shows the various interfaces used to connect to the PCRF. The Sd interface to the traffic detection function can also be labeled as deep packet inspection. This function is what provides details about the destination for a session so that policy can be applied based on the destination application (such as YouTube, or some other content provider network).

The Sy interface is a fairly new addition to the standards, providing a direct connection from the PCRF to the online charging system (OCS). There is a lot of dialog between online charging and the PCRF, making the combination of the PCRF, the various enforcement points, and the OCS one of the largest generators of Diameter signaling traffic.

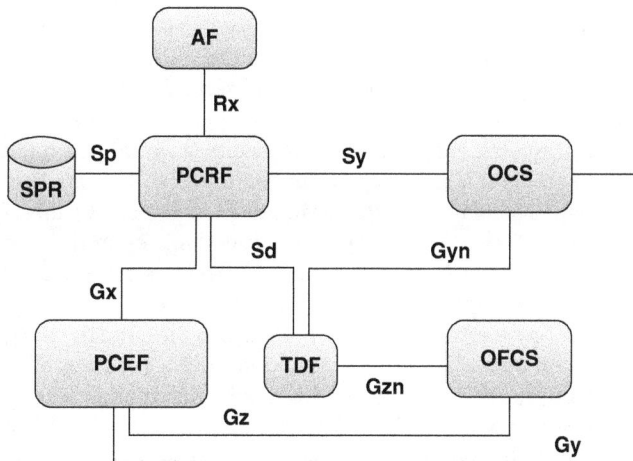

FIGURE 7.1 Policy Charging Control (PCC) interfaces.

Introduction to Policy Management

When a service provider creates a policy, they define the policy in the PCRF through provisioning. They can define policies that are then applied to every subscriber, with a number of key attributes:

- Time of day
- Type of device
- Service being requested
- Application or Internet site being accessed
- Subscription plan or other subscription criteria

Each rule is based on a trigger of some type, such as a device accessing the network, or a device attempting to connect to an Internet service (such as YouTube or social media). As the data flow makes its way through the network, each of the enforcement points will query the PCRF to determine how the session should be treated.

The PCRF checks the subscriber profile stored in the UDR/SPR to determine how traffic for this user is to be handled. The subscriber may have special privileges that other subscribers do not have, based on their data plan or other factors. There are a number of attributes that can be applied, depending on the enforcement point that is requesting information.

The session type is also important, because the PCRF may need to decide the quality of service (QoS) to be provided for the session. For example, if the subscriber is connecting to a video service and the provider of the video service is sponsoring the connection, then the subscriber could get a higher QoS, depending on the arrangement between the service provider and the data sponsor. The sponsor subsidizes the higher bandwidth for a group of subscribers as defined in the PCRF and UDR/SPR.

The PCRF may also look at the type of device to determine the best option for the video format. It would not be efficient to send a high definition video to a device that does not support high definition, so the video may be modified through video optimization before sending to the device. The PCRF plays an important role in the video optimization.

When working with charging, the PCRF will query the subscriber account in the charging system to determine how much "credit" the subscriber has on account. Think of credit as quota. The charging platform will send the PCRF an amount (either in dollars or in units of time/volume). Quota management is one of the most popular use cases for policy. Subscribers are allowed a specific amount of data quota, and when that quota has been reached, the subscriber must "top-up" their account.

The rules provisioned by the PCRF define the actions to be taken by the enforcement point. The PCRF acts as a middleman between charging and the enforcement point. If the session is to be terminated by the enforcement point because quota has been exhausted, the PCRF applies this in the rules provisioned for the subscriber session.

The Flow-Information AVP, the AVP describes the state for the gate (open or closed) in the enforcement point. The gate is what determines if data is allowed to pass through or not. For example, to restrict a mobile device from accessing data sessions during specific hours, a rule would mark the gate closed for that traffic during the specified hours at the node. The mobile device would not be able to establish a data session during that time (parental controls, for example). The traffic is simply discarded if the gate is closed, or allowed to pass through if the gate is open. This can be applied on the TDF via the Sd interface for applications as well.

This concept of the PCRF providing provisioning rules is important, because it must be understood that the PCRF does not actually enforce the rules. It simply defines them and provisions the nodes when a session is being requested.

Policy Use Cases

There are many different use cases for policy. Typically, a service provider will start with traffic management use cases. For example, to manage the amount of bandwidth that is consumed by subscribers, the service provider uses fair usage policy. In many networks, a very small percentage of subscribers (3 percent in several cases) consume more than 80 percent of the total network bandwidth. This means the remaining 97 percent of the subscribers may receive poor quality connections because there is not enough bandwidth left for their sessions.

Fair usage ensures subscribers are not using more than the rest of the subscribers, and actually throttles traffic from bandwidth hungry subscribers.

As you can see in Fig. 7.1, the PCRF interacts with each of the enforcement points in the bearer path individually. Each enforcement point may provide different information. For example, the SGSN does not know the URL a subscriber is requesting, but the TDF does know this information, and will provide this to the PCRF so it can provision based on the website.

Another use case is quota management. Prepaid subscribers have a finite amount of credit on their account and they are only granted services as long as they have credit on account. The PCRF maintains a dialog with charging for each subscriber session, ensuring the subscriber does not exceed their credit limit. When they reach a certain threshold, the PCRF may facilitate the sending of a text message or other alert so the subscriber can top up their account.

Sponsored data sessions have become more and more commonplace and a great revenue source for service providers. The sponsor (usually the content provider) subsidizes the session to ensure there is enough bandwidth provided for the session for a high-quality experience. The PCRF rules make sure that a sponsored data session is not charged to the subscriber, and apply appropriate charges to the data sponsor.

There are many more use cases we could discuss, but in the interest of brevity we will not attempt to cover them all here. The PCRF provides a capability that service providers need to enhance the subscribers' experience and to optimize the packet services they provide. Note that policy also applies to M2M and Internet of Things (IoT) traffic as well.

Rx Interface

The first interface we will talk about is the Rx interface. The Rx interface connects the PCRF with an application function (AF) in the IMS. An AF is typically the CSCF. Since the first access a mobile device has with the IMS is the Proxy-CSCF (P-CSCF), this is the example we will use in this section, rather than always referring to the application function. However, you will see AVPs that reference the AF.

The P-CSCF exchanges information about an IMS session with the PCRF. The IMS is used to support voice calls in the Long Term Evolution (LTE) network. This is also referred to as Voice over LTE (VoLTE). The P-CSCF can be in the same network, or in another network, using the Rx interface across network boundaries.

Remember that the mobile device is accessing the IMS via the packet core, which means it must first connect through the SGSN, and the PDN gateway. There will be an interaction therefore between the enforcement points and the PCRF during Rx sessions using the Gx interface.

One example of this is QoS. When the mobile device is connecting to the IMS, the SGSN and the PDN Gateway will send the PCRF information about the QoS to be provided by the enforcement point, as well as any charging considerations. This is then passed to the CSCF via the Rx interface during the information exchange between the PCRF and the CSCF.

Rx Procedures

When a node is connecting with the PCRF to establish a session, it must first negotiate its capabilities to ensure that the PCRF it is trying to connect with can support all of the applications it may need. During the capabilities exchange, the network elements will share the applications they support. The Rx Application-ID is 16777236, and the Vendor-ID is 10415 (3GPP). These are advertised in the CER/CEA using the Auth-Application-ID and the Vendor-Specific-Application-ID AVPs.

The mobile device connects to the P-CSCF that will then send the AA-Request (AAR) command to the PCRF requesting a session. The AAR will include the IP address of the mobile device in the Framed-IP-Address, or the Framed-IPv6-Prefix AVPs, depending on the IP address format supported by the mobile device.

Remember that no media is sent on the Rx interface, so there is no media in the AAR command. The media is sent using the Session Initiation Protocol (SIP). If the mobile device is making a voice connection using the circuit-switched network, then packet services are not used and there is no Rx session. The Rx is only used to support a VoLTE session, or other multimedia sessions in the IMS.

The CSCF will provide information about the service being provided in the Media-Component-Description AVP, including the priority to be applied to the session. Note that there is a separate negotiation going on between the mobile device and the IMS in the SIP domain. If the CSCF has accepted the media session as requested from the mobile device, the CSCF will send the PCRF the Service-Info-Status AVP with the value of FINAL_SERVICE_INFORMATION. This means the CSCF has completed the SIP negotiations for the session and has agreed to provide the service.

The PCRF can then use the received information for provisioning the enforcement points in the packet core regarding the service being provided, the QoS to be provided, and any other rules to be applied to the session.

Priority is provisioned in two places. First, the radio access network (RAN) must apply priority at the radio to ensure the mobile device is given priority over other devices trying to access the same cell and radio. Once the mobile device has reached the packet core, priority is applied again to ensure the device is given priority to the packet services.

Priority for a session is identified in two different ways. The Reservation-Priority AVP can be used in the AAR to indicate the priority for the overall session. A session may consist of multiple IP flows, and so in this case the priority would be given to all of the IP flows for the given session.

At the same time, the Reservation-Priority AVP can be sent in the Media-Component-Description AVP to indicate the priority to be provisioned for each IP flow. It is acceptable to use the Reservation-Priority AVP in both places.

The PCRF sends an authentication, authorization, accounting (AAA) command in response to the AAR, while provisioning the enforcement points via the Gx interface. If the PCRF is unable to provision the requested QoS in the enforcement points for any reason, the PCRF will send the RAR command to the CSCF using the Specific-Action AVP set to the value of INDICATION_OF_FAILED_RESOURCES_ALLOCATION.

The CSCF can send updates and changes for a session to the PCRF throughout the lifetime of the session. However, the CSCF must have first received the AAA command in response to the initial AAR prior to sending any updates to the PCRF. The CSFC cannot send updates to a session it has not received a response for.

To terminate an Rx session, the CSCF sends the Session-Termination-Request (STR) command to the PCRF. The PCRF then in turn will request the release of any connections in the packet core with the enforcement points by sending an STR to each of the enforcement points via the Gx interface.

One exception to this rule is when sponsored data is being supported for a session. The PCRF will delay sending the STA command to the CSCF in response to the STR until any threshold that may have been sent for the sponsored session has expired. The PCRF then sends the User-Service-Unit AVP within the Sponsored-Connectivity-Data AVP to indicate how much data was consumed for the sponsored session.

The CSCF can request notification from the PCRF anytime there is a change in the bearer connections for a session, or for usage on a periodic basis. The CSCF sends the PCRF this request in the AAR, and includes the threshold to be applied for the reporting intervals. The PCRF uses this threshold for determining when it needs to send notification to the CSCF.

When sending usage information, the PCRF sends the RAR command with the Specific-Action AVP set to the value of USAGE_REPORT. The PCRF does not track usage itself, since the bearer does not connect through the PCRF. The usage information must come from the enforcement points in the packet core, via the Gx interface.

If requesting notification for a change in the connection status, the CSCF will include the mobile device IP address in the AAR. The Specific-Action AVP is used with the value set to INDICATION_OF_LOSS_OF_BEARER and/or INDICATION_OF_RELEASE_OF_BEARER. The Media-Component-Description AVP is sent containing the Media-Sub-Component AVP with the Flow-Usage AVP value of AF_SIGNALING.

When the CSCF needs location information for a specific mobile device, the CSCF requests the PCRF to send this data using the Specific-Action AVP set to the value of ACCESS_NETWORK_INFO_REPORT. The Required-Access-Info AVP will identify the type of information that is being requested. This could include the location, the type of access method being used (WiFi, 3G, 4G, etc.), or the time zone for the mobile devices current location.

The PCRF sends this information using the AAA command. If the enforcement point does not support network access notification, the PCRF will know this because it will have already exchanged the Supported-Features AVP with the enforcement point during CER/CEA. The PCRF will send a response back to the CSCF using the NetLoc-Access-Support AVP set to the value of NETLOC_ACCESS_NOT_SUPPORTED.

When the data connection in the packet core has been terminated, the PCRF notifies the CSCF via the Rx interface by sending the Abort-Session-Request (ASR) command.

The CSCF returns the ASA command followed by the Session-Termination-Request (STR) command to terminate the Rx session.

Gx Interface

The Gx interface connects the PCRF with policy enforcement points. These are found in the packet core of 3G and 4G networks. The serving GPRS support node (SGSN), gateway GPRS support node (GGSN), and PDN Gateway are all considered as enforcement points, supported by the Gx interface. Note that the Rx and the Gx may work together when delivering a service to the customer, as will the Sd interface.

Gx Session Establishment

When a mobile device starts a data session in the wireless network, the user data makes its way to the packet network, where it will be routed to an enforcement point. We are not going to go through the traffic routing for the sake of brevity and focus. When it reaches an enforcement point, the enforcement point will send a CCR command to the PCRF using the Gx interface.

This is one of two ways that the PCRF communicates with the enforcement point. The enforcement point uses the CCR command when the device initiates a data session. When the mobile device initiates an IMS session, the PCRF uses the RAR command (more on that later).

Because this is the first communication to the PCRF, the enforcement point needs to mark this command as an initial request. This is done using the CC-Request-Type AVP value of INITIAL_REQUEST.

If a mobile device makes an emergency call, session establishment is a bit different. The enforcement point sends a CCR command to the PCRF containing the Called-Station-ID AVP identifying the Emergency APN. The PCRF stores a list of emergency APNs it is associated with and checks the UDR/SPR for rules to be applied for the specified APN.

The PCRF may also receive the IMSI of the subscriber in the Subscription-ID AVP, but it is not necessary. If no IMSI is provided, the enforcement point includes the IMEI for the mobile device using the User-Equipment-Info AVP. The PCRF returns the CCA with the rule to be provisioned for the session, limiting access to just the emergency APN and no other applications.

QoS is also set for the session based on the value set by the UDR/SPR. The session priority is set using the Priority-Level AVP within the rule. The service provider should have established various priority levels for various types of sessions in the network with emergency having the highest priority.

Since this session is an emergency voice call, the PCRF will receive another session request from the IMS when the packet core routes the traffic to the IMS. The packetized voice will be routed to the P-CSCF in the IMS, which in turn will send a session request to the PCRF. The PCRF will need to send the enforcement point rules for this session based on the IMS connection, which means there will need to be an association made between the packet core session and the IMS session. This is where the device IP address, IMSI, and any other device/subscriber data become important.

A Gx session can be modified at any time by sending another CCR command to the PCRF, but only after the enforcement point has received a CCA from the PCRF for the initial session request. There are three conditions where a session can be modified at the Gx interface:

- The IP-CAN is being established or terminated
- Modification of resources for a session is being requested
- An event trigger has been met

The enforcement point sends the new CCR with the CC-Request-Type AVP value of UPDATE_REQUEST. If the session has already been established and is being modified, the Event-Trigger AVP is also sent to identify what triggered the modification.

The mobile device can be the cause of a session modification if it requests changes to an established session, resulting in the change to resources for the session. In this case, the CC-Request-Type value is RESOURCE_MODIFICATION_REQUEST. The Packet-Filter-Operation AVP is included for each additional packet filter being requested by the enforcement point. The value for the Packet-Filter-Operation AVP is set to ADDITION for adding new packet filters, or MODIFICATION if simply modifying existing packet filters. The value is set to DELETION if packet filters are being deleted.

The mobile device, the enforcement point, or the PCRF can terminate IP-CAN sessions. If the enforcement point terminates an IP-CAN session, it will notify the PCRF using the CCR command with the CC-Request-Type AVP value TERMINATION_REQUEST.

The PCRF can terminate a session two ways. If the session is a data session, the PCRF can terminate the IP-CAN session by sending the Session-Release-Cause AVP in the CCA. If the PCRF is terminating an IMS session, it will use the RAR command and the Session-Release-Cause AVP.

If the enforcement point detects that something has caused the default bearer channel to be terminated, or if it is unable to establish a dedicated bearer channel for a requested user session, it sends the CCR command to the PCRF. Any rules provisioned for the bearer channel are first removed. The enforcement point sends the CC-Request-Type AVP set to UPDATE_REQUEST including the Charging-Rule-Report AVP listing all of the rules that have been disabled. The enforcement point also sends the PCC-Rule-Status AVP value INACTIVE, with the Failure-Code value of RESOURCE_ALLOCATION_FAILURE.

If for any reason the enforcement point is unable to install or activate a rule, it notifies the PCRF by sending the PCC-Rule-Status value of INACTIVE. For rules pushed to the enforcement point (sent by the PCRF) the enforcement point sends the CCR command to the PCRF with the Rule-Failure-Code AVP notifying the PCRF of the failure. For rules that were pushed to the enforcement point, the enforcement point uses the Re-Auth-Answer (RAA) to notify the PCRF. The failed rules are identified in the Charging-Rule-Name or Charging-Rule-Base-Name AVP. The failure cause is identified in the Rule-Failure-Code AVP.

If the enforcement point is no longer able to support a rule, the enforcement point notifies the PCRF by sending the Charging-Rule-Report AVP, along with the Rule-Failure-Code identifying the cause of the failure. The PCC-Rule-Status is marked as INACTIVE.

In the event the PCRF cannot send rules to the enforcement point, it will respond to a request by sending the Experimental-Result-Code DIAMETER_ERROR_INITIAL_PARAMETERS. The enforcement point will then reject the request for an IP-CAN connection of modification.

The PCRF could also be in the process of provisioning the enforcement point with new rules when it receives a request for modifications to IP filters already being modified. This represents a conflict in the PCRF, which will respond with the Experimental-Result-Code DIAMETER_ERROR_CONFLICTING_REQUEST.

The PCRF uses the Event-Trigger AVP to remove all triggers from the enforcement point by setting the value to this AVP to NO_EVENT_TRIGGERS. This removes any triggers from the enforcement point.

Gx Procedures

The PCRF provisions rules in the enforcement point by sending the CCA command via the Gx interface. It may also send rules in the RAR command, if rules are associated with IMS sessions. We will discuss this procedure later.

The enforcement point requests rules from the PCRF by sending the CCR command. The rules are then sent by the PCRF using the CCA as a response. If the PCRF is sending rules to the enforcement point unsolicited (as would be the case for an IMS session or an internal trigger in the PCRF, for example), the PCRF uses the RAR command. Note that the PCRF would have received a request from the IMS via the Rx interface prior to sending rules in an RAR to the enforcement point.

Rules can also be preprovisioned in the enforcement point. These rules are then activated or deactivated by the PCRF. They can also be modified by the PCRF. Each rule (or group of rules) is assigned a name that can be referenced by the PCRF when it needs to activate, deactivate, or modify the rule. The name is then referenced in the Charging-Rule-Name AVP. A group of rules is identified using the Charging-Rules-Base-Name AVP.

The Charging-Rule-Install or the Charging-Rule-Remove AVPs indicate the action to be taken for the specified rule, or group of rules. These AVPs can be sent in one command, referencing multiple rules rather than sending many commands. It is these AVPs that provision a rule in the enforcement point.

If the PCRF wishes to provision a rule at the request of the enforcement point (in other words the enforcement point has sent the CCR command requesting a rule), it sends the CCA command with the Charging-Rule-Install AVP and the rule described in the Charging-Rule-Definition AVP. The Charging-Rule-Definition is included in the grouped Charging-Rule-Install AVP. This is also used to modify rules that are preprovisioned in the enforcement point.

When removing rules from the enforcement point, the PCRF sends the Charging-Rule-Remove AVP within the CCA or the RAR command. The rule to be removed is identified using the Charging-Rule-Name or the Charging-Rule-Base-Name AVPs.

The PCRF can also change the address of charging, redirecting charging records to a different system. The Charging-Information AVP containing the address of the offline charging system (OFCS) or the OCS is used when a rule is applied. Even if the charging address has already been provisioned in the enforcement point, there are times when charging needs to be redirected based on the traffic type. The charging address is sent when the initial rule is installed.

If the PCRF does not know the access network charging identifier to be used for IMS sessions, it can request the enforcement point to send it. The PCRF does this by sending the Event-Trigger AVP in CCA or RAR command with the value CHARGING_CORRE-LATION_EXCHANGE. This is sent along with the Charging-Rule-Install AVP containing the Charging-Correlation-Indicator AVP value of CHARGING_IDENTIFIER_REQUIRED.

The enforcement point would then send the information back in the Access-Network-Charging-Identifier-Gx AVP.

There may also be information available regarding the type of access technology that was used by the mobile device to connect to the network. This could be important to the PCRF, because there are use cases where the traffic is handled differently depending on how the mobile device gained access. For example, if the device is within close proximity of WiFi operated by the service provider and the mobile device is connecting to a video streaming application, the service provider may choose to offload this traffic to WiFi rather than allow the device to use the 4G networks.

This is done by the PCRF. The IP-CAN-Type AVP provides the information about the type of access (xDSL, WiMAX, 3GPP-GPRS, etc.). The RAT-Type AVP identifies the radio technology (3G, 4G, WiFi) if accessed via the wireless network. Both of these together are important for the PCRF to understand the rules to be applied.

The mobile device is assigned an IP address when it connects to the network. The PCRF uses this information to be able to reference a specific device when managing connectivity, QoS, and other rules specific to the mobile device. The Framed-IP-Address or the Framed-IPv6-Prefix AVPs provide the IP address assigned to the mobile device and are included in the CCR (when available).

The type of equipment (make, model, and software version, for example) can be used in a rule to govern the type of QoS, the type of applications allowed, and even the type of access allowed for a specific subscriber. For example, a subscriber may be connecting to a video streaming service using a device that does not support high definition. By identifying the type of device, the network can use video optimization to modify the video streaming to the device that will reduce the amount of bandwidth required to send the video to the device, and provide an optimal experience for the subscriber. This is governed by the PCRF when the User-Equipment-Info AVP is included providing it information about the mobile device.

Quality of Service

There are many different ways in which the PCRF can provision the enforcement point for QoS. QoS can be assigned to an entire IP-CAN, to an APN, to a specific access type, or even the type of device being used. The PCRF uses the CCA command to send QoS parameters to the enforcement point upon request, or in the case of an IMS session, the QoS is pushed to the enforcement point using the RAR command. The QoS requirements are sent using the QoS-Information AVP. This AVP includes the QCI, ARP and bit rates required for the session or to be applied for the IP-CAN. The PCRF identifies the bearer channel where the rule is to be applied using the Bearer-Identifier AVP.

The quality class identifier (QCI) is used to identify what QoS is to be used for a session. There are nine QCIs defined for LTE networks. The highest bitrate allowed is defined in the QCI, and applied to the rule or IP-CAN.

QoS can also be assigned to a specific APN, using the APN-Aggregate-Max-Bitrate-UL or the APN-Aggregate-Max-Bitrate-DL AVP sent in either the CCA command or the RAR command. This type of rule is different than others in that it is applied to all sessions for the specified APN. In other words, the rule is applied to every IP-CAN and radio access type for the specified APN.

An alternative approach for assigning QoS is to provision QoS on each IP-CAN accessing a specific APN, rather than a single rule applied to all IP-CANs connecting to the APN. This allows the PCRF to define QoS based on the radio access type, and the device type for ach APN.

Finally, the QoS can be assigned for the default bearer channel by sending QoS parameters in the Default-EPS-Bearer-QoS AVP. The PCRF sends the QoS-Class-Identifier and the Allocation-Retention-Priority AVPs within the Default-EPS-Bearer-QoS AVP. These are sent in either the CCA or the RAR commands.

If the enforcement point is unable to modify the QoS for the default bearer, it will send a new CCR containing the Event-Trigger AVP set to the value of DEFAULT-EPS-BEARER-QOS _MODIFICATION_FAILURE. The enforcement point maintains the previous QoS provisioned for the default bearer, sending this value to the PCRF in the Allocation-Retention-Priority and QoS-Class-Identifier AVPs within the Default-EPS-Bearer-QoS AVP.

Some rules may need to be installed or activated at a specific time (such as the rule we mentioned earlier regarding parental controls). The PCRF uses the Rule-Activation-Time or the Rule-Deactivation-Time AVP within the Charging-Rule-Install AVP to specify the times a rule is to be activated or deactivated. Timers cannot be applied to rules that change the QoS or the IP filters in the enforcement point.

If a rule is to be changed at a specific time-of-day (e.g., a rule may need to be set to allow specific types of sessions during the day) the PCRF uses the Event-Trigger AVP value of REVALIDATION_TIMEOUT. The time-of-day is included in the Revalidation-Timeout AVP. This sets a timer in the enforcement point, and when the timer expires, the enforcement point will send a new CCR to the PCRF to request new rules.

When the PCRF responds with the CCA command, it includes a new timer value by sending another Revalidation-Timeout AVP. If the PCRF needs to stop the timer in the enforcement point, it sends the Event-Trigger AVP value of REVALIDATION_TIMEOUT.

Usage Monitoring and Notification

The PCRF can request the enforcement point to send notification when resources specific to a rule have been allocated. This is done by the PCRF sending the Charging-Rule-Install AVP with the Resource-Allocation-Notification AVP set to the value of ENABLE_NOTIFICATION in a CCA command. The enforcement point sends a new CCR to the PCRF containing the Event-Trigger AVP set to the value of SUCCESSFUL_RESOURCE_ALLOCATION.

Some rules may require information regarding usage from the enforcement point. Usage monitoring is managed by the enforcement point upon request from the PCRF. For example, a prepaid account would need to be blocked of network privileges when the subscriber runs out of credit. A monitoring key is used to identify the monitoring instance. The PCRF assigns the Monitoring-Key AVP within the Charging-Rule-Install AVP.

The usage information can be provisioned on a specific IP-CAN or group of IP-CANs or for a specific mobile device. Usage monitoring is based on volume or time, or both. The PCRF will define how much volume/credit is allowed, and the enforcement point reports back to the PCRF when the allotment has been exhausted.

To request usage monitoring, the PCRF sends the Event-Trigger AVP value of USAGE_REPORT. This is sent either in the CCA or the RAR command, depending on the type of session. The parameters for the usage monitoring can be sent when the rule is provisioned or when modifying an existing session.

Monitoring is assigned either to a rule or to an IP-CAN session. The Usage-Monitoring AVP defines how the usage monitoring is to be applied:

- SESSION_RULE
- PCC_RULE_LEVEL

Only one usage monitoring instance can be provisioned for an IP-CAN, but there can be multiple instances for a rule. To exclude some data flows from monitoring on an IP-CAN, the Monitoring-Flags AVP with the bit set to zero is sent during rule provisioning.

Threshold information for a specific Monitoring-Key is provided in the Usage-Monitoring-Information AVP. The Granted-Service-Unit AVP identifies the thresholds to be provisioned for monitoring. The thresholds can be identified for

- Total volume
- Uplink volume
- Downlink volume
- Uplink and downlink volume
- Time

The CC-Total-Octets, CC-Input-Octets, CC-Output-Octets, and CC-Time AVPs are provided in the Granted-Service-Unit AVP for these thresholds. The enforcement point sends usage information using these same AVPs in the CCR command back to the PCRF, but they are contained within the Used-Service-Unit AVP that is part of the Usage-Monitoring-Information AVP. Each monitoring key is represented by its own Usage-Monitoring-Information AVP.

The time at which monitoring begins and ends is also defined by the PCRF when requesting usage monitoring. The PCRF uses the Monitoring-Time AVP to define the start and stop times for each monitoring key. One set of Granted-Service-Unit AVPs and Usage-Monitoring-Information AVPs is sent for the start time and another for the stop time. When the monitoring period stops, the enforcement point will send a report to the enforcement point regarding usage using the CCR command.

The PCRF will then respond to the report sending a CCA with new monitoring usage thresholds and times if the monitoring is to continue. If usage monitoring is not going to be continued, the PCRF sends nothing in the CCA.

If the enforcement point sends a CCR command to the PCRF without usage information, the PCRF can request usage information by sending the CCA command with Usage-Monitoring-Information AVP containing the Usage-Monitoring-Report AVP set to the value of USAGE_MONITORING_REPORT_REQUIRED. If the PCRF does not include the Monitoring-Key AVP to identify a specific monitoring key, the enforcement point sends usage information for all the monitoring keys.

When there is no activity detected by a monitoring key on an IP-CAN or session, the enforcement point will stop monitoring after a predetermined time. This time is provisioned in the Monitoring-Key by the PCRF using the Quota-Consumption-Time AVP. When a packet is received again and detected by the monitoring key, monitoring continues.

Sd interface

Another use case for PCRF is the ability to redirect traffic to another network resource. For example, a subscriber who is looking to establish a data connection could be redirected to a local WiFi network sponsored by the service provider. This requires a number of things in the network to support, but the traffic detection function (TDF) is one of those requirements.

The TDF is responsible for detecting the type of traffic. It uses rules to determine if the traffic should be redirected, or receive special treatment. The TDF communicates with the PCRF using the Sd interface.

The TDF is different from the enforcement point even though they both provide policy enforcement. The TDF focuses on applications rather than a specific data flow. The enforcement point does not have any knowledge of the application being routed to, because it is only looking at the data flow itself, and not into the packet. The TDF uses deep packet inspection to identify the application addresses (such as IP address or URL).

A good example of this might be a subscriber connecting to a video source. The video source is going to require additional bandwidth (QoS) that may not be provided at the enforcement point, which only sees a data connection. The TDF sends a rules request to the PCRF when it receives the packet, and the PCRF provisions the additional rules to be applied at the TDF.

Sd Procedures

Rules are provisioned on the TDF the same way they are provisioned at the enforcement point. The PCRF can either push rules to the TDF or send rules when requested (when pulled). In the standards, you will hear rules at this part of the network referred to as application detection and control (ADC) rules.

The PCRF will subscribe to the TDF for reporting, using the event triggers to define when the TDF should report to the PCRF. The event triggers will identify traffic for an application has started or stopped. The TDF reports usage relevant to the applications to the PCRF using usage monitoring.

Since traffic will travel through both the enforcement point and the TDF, the PCRF is able to link Sd sessions with Gx sessions. The PCRF receives information from the Gx as well as the Sp and Sd interfaces that can be applied to the rules provisioned in the TDF. The SPR would provide subscriber data, while the enforcement point would provide information about the connection such as the APN, data flow, and type of network access used.

The PCRF matches the IP addresses provided by the TDF with IP addresses received from the enforcement point on the Gx interface. If there is a match, then the two sessions are linked. PDN information from the Called-Station-ID AVP can also be used for this linkage.

Traffic can also be redirected by the TDF. For example, access network discovery and selection function (ANDSF) requires a TDF for determining the type of traffic, and the PCRF provides instructions on where to redirect the traffic (such as a WiFi network).

The PCRF uses the Redirect-Information AVP contained in the Charging-Rule-Definition AVP for providing instructions to the TDF for traffic redirection.

The address to be used for redirecting traffic can be preprovisioned in the TDF or sent by the PCRF. The address is provided by the PCRF using the Redirect-Server-Address AVP in the rule. This overrides any preprovisioned address in the TDF. The PCRF can also disable redirection of traffic by sending the Redirect-Support value of REDIRECTION_DISABLED.

Only the uplink traffic is redirected. The Redirect-Address-Type AVP identifies the type of address being provided in the Redirect-Address-Server AVP. It can be an IP address or URL.

Other controls provided at the TDF include

- ADC rule identifier
- Application identifier
- Precedence indicator
- Charging key and charging parameters
- Monitoring key
- Gate status
- Upload and download maximum bit rates
- Download DSCP value

The ADC rule identifier identifies each unique rule provisioned in the TDF. This is important for identifying rules to be activated or deactivated, just as we saw with the Gx interface and the enforcement point.

Precedence determines which rules get applied in the event the PCRF sends dynamic rules to the TDF and there are already preconfigured rules defined, with the same ADC rule identifier. The application identifier is used to identify the application for which the rule applies. Dynamic rule always take precedence over preconfigured rules with the same ADC identifier.

When the PCRF is provisioning a rule at the TDF, it can also request detection of traffic to a specific application. When the PCRF sends the Charging-Rule-Install AVP, it includes the TDF-Application-Identifier to provide the information about the application to be detected. This is sent as part of the Charging-Rule-Definition AVP.

If the TDF is already provisioned with a rule for this and the PCRF has activated the rule, the enforcement point uses the Application-Detection-Information AVP in the CCR command. The TDF-Application-Identifier is also sent within the Application-Detection-Identifier AVP.

The enforcement point will send the CCR command to the PCRF containing the Event-Trigger AVP set to APPLICATION_START when traffic is detected for an application. This indicates usage monitoring has begun for the application traffic. The enforcement point also sends the Flow-Information AVP that contains the Flow-Description and Flow-Direction AVPs. When the usage monitoring stops, the enforcement point sends the CCR command with the Event-Trigger AVP set to APPLICATION_STOP.

Charging for applications is controlled by the TDF by identifying the charging address to be used for specific applications so specific sessions. The TDF acts as a traffic gate at the user plane, routing traffic or discarding traffic based on the rules provisioned by the

PCRF. The subscriber's available credit is used to determine if the traffic can be routed or should be discarded. The Flow-Status AVP defines the gate to be applied for a rule.

If the PCRF is defining the QoS to be applied for a specific application, it uses the QoS-Information AVP. The Max-Requested-Bandwidth-UL and the Max-Requested-Bandwidth-DL AVP provide the maximum amount of bandwidth to be allowed for the specified application.

When requesting usage monitoring from the TDF, the PCRF sends an Event-Trigger value of USAGE_REPORT in the TDF-Session-Request (TSR), CCR, CCA, or RAA command. The TDF returns usage information using the CCR command. The Monitoring-Key sent in the ADC-Rule-Definition identifies the specific monitoring instance in the TDF.

The Monitoring-Key can be applied to one or multiple application sessions, or to all detected traffic for a specified TDF session. If all of the sessions are being monitored, the TDF can be instructed to exclude some applications. This is useful when "zero-rating" certain application traffic. For example if the service provider wants to provide connections to a specific content provider at no charge, they can exclude usage monitoring to the content provider. The Usage-Monitoring-Level AVP is used here in the same way it is used with the enforcement point, indicating either SESSION_LEVEL or ADC_RULE_LEVEL.

Monitoring based on the volume of traffic and the time accessing an application can be applied at the same time. The volume thresholds to be applied are sent in using CC-Total-Octets, CC-Input-Octets or CC-Output-Octets AVPs and/or time threshold in the CC-Time AVP of the Granted-Service-Unit AVP.

The Monitoring-Time AVP requires two Granted-Service-Unit AVPs sent in the Usage-Monitoring-Information AVP for each monitoring key. The first defines the threshold to be applied before monitoring is to begin (in other words, how much time is to elapse once the traffic is detected before usage monitoring is applied). The second threshold is used to identify the thresholds after the traffic is detected.

An inactivity time can also be defined using the Quota-Consumption-Time AVP. This value is used to stop monitoring when there is no activity detected for the specified time interval. Once traffic resumes for the specified monitoring key, usage recording resumes.

When the TDF is ready to report usage, it sends the CCR command to the PCRF with the accumulated usage information. The Usage-Monitoring-Information AVP contains this information. The TDF can report one time, or it can report in intervals, meaning the PCRF will receive multiple CCRs for the same traffic. Each time the TDF reports the accumulated usage since the previous report.

The actual usage consumed is provided using the Used-Service-Unit AVP, which contains the CC-Total-Octets, CC-Input-Octets, or CC-Output-Octets AVPs. When reporting time, the TDF uses the CC-Time AVP.

Commands Used on Rx, Gx, and Sd

The commands used to connect with the PCRF can be the same at each interface, with some nuances. For example, the CCR is used across several of the interfaces, but the content of the CCR may be different depending on which interface is being used. Rather than try and provide the same command several times, we only show the command once here with all the possible AVPs.

Credit Check Request/Answer

The Credit Check Request (CCR) command is the most commonly used command between the enforcement point and the PCRF. The enforcement point uses the CCR for requesting rules for a specific session. The enforcement point also uses the CCR for reporting events in the network such as routing of sessions or termination of a bearer. The CCR is always sent from enforcement point to the PCRF.

<CC-Request>::=< Diameter Header: 272, REQ, PXY>

<Session-Id>

{Auth-Application-Id}

{Origin-Host}

{Origin-Realm}

{Destination-Realm}

{CC-Request-Type}

{CC-Request-Number}

[Credit-Management-Status]

[Destination-Host]

[Origin-State-ID]

*[Subscription-ID]

[OC-Supported-Features]

*[Supported-Features]

[TDF-Information]

[Network-Request-Support]

*[Packet-Filter-Information]

[Packet-Filter-Operation]

[Bearer-Identifier]

[Bearer-Operation]

[Dynamic-Address-Flag]

[Dynamic-Address-Flag-Extension]

[PDN-Connection-Charging-ID]

[Framed-IP-Address]

[Framed-IPv6-Prefix]

[IP-CAN-Type]

[3GPP-RAT-Type]

[AN-Trusted]

[RAT-Type]

[Termination-Cause]

[User-Equipment-Info]

[QoS-Information]

[QoS-Negotiation]

[QoS-Upgrade]

[Default-EPS-Bearer-QoS]

[Default-QoS-Information]

0*2[AN-GW-Address]

[AN-GW-Status]

[3GPP-SGSN-MCC-MNC]

[3GPP-SGSN-Address]

[3GPP-SGSN-IPv6-Address]

[3GPP-GGSN-Address]

[3GPP-GGSN-IPv6-Address]

[3GPP-Selection-Mode]

[RAI]

[3GPP-User-Location-Info]

[User-Location-Info-Time]

[User-CSG-Information]

[TWAN-Identifier]

[3GPP-MS-TimeZone]

*[RAN-NAS-Release-Cause]

[3GPP-Charging-Characteristics]

[Called-Station-ID]

[PDN-Connection-ID]

[Bearer-Usage]

[Online]

[Offline]

*[TFT-Packet-Filter-Information]

*[Charging-Rule-Report]

*[Application-Detection-Information]

*[Event-Trigger]

[Event-Report-Indication]

[Access-Network-Charging-Address]

*[Access-Network-Charging-Identifier-Gx]

*[CoA-Information]

*[Usage-Monitoring-Information]

[Routing-Rule-Install]

[Routing-Rule-Remove]

[HeNB-Local-IP-Address]

[UE-Local-IP-Address]

[UDP-Source-Port]

[Presence-Reporting-Area-Information]

[Logical-Access-ID]

[Physical-Access-ID]

*[Proxy-Info]

*[Route-Record]

*[AVP]

The PCRF responds to the CCR by sending the Check Credit Answer (CCA) command. The PCRF may also include rules to be provisioned at the enforcement point in the CCA.

<CC-Answer>::=<Diameter Header: 272, PXY>

<Session-Id>

{Auth-Application-ID}

{Origin-Host}

{Origin-Realm}

[Result-Code]

[Experimental-Result]

{ CC-Request-Type}

{ CC-Request-Number}

[OC-Supported-Features]

[OC-OLR]

*[Supported-Features]

[Bearer-Control-Mode]

*[Event-Trigger]

[Event-Report-Indication]

[Origin-State-Id]

*[Redirect-Host]

[Redirect-Host-Usage]

[Redirect-Max-Cache-Time]

*[Charging-Rule-Remove]

*[Charging-Rule-Install]

[Charging-Information]

[Online]

[Offline]

*[QoS-Information]

[Revalidation-Time]

[Default-EPS-Bearer-QoS]

[Default-QoS-Information]

[Bearer-Usage]

*[Usage-Monitoring-Information]

*[CSG-Information-Reporting]

[User-CSG-Information]

[Presence-Reporting-Area-Information]

[Session-Release-Cause]

[Error-Message]

[Error-Reporting-Host]

*[Failed-AVP]

*[Proxy-Info]

*[Route-Record]

*[AVP]

Re-Auth-Request/Answer

The Re-Auth-Request (RAR) command is used between the PCRF and the enforcement point when the PCRF needs to push rules to the enforcement point. This usually is the case when an IMS session is established, and the P-CSCF sends session information to the PCRF. The PCRF then pushes rules for the IMS session at the packet core using the RAR command on the Gx.

<RAR>::=< Diameter Header: 258, REQ, PXY>

<Session-ID>

{Auth-Application-ID}

{Origin-Host}

{Origin-Realm}

{Destination-Realm}

{Destination-Host}

{Re-Auth-Request-Type}

[Session-Release-Cause]

[Origin-State-ID]

[OC-Supported-Features]

*[Event-Trigger]

[Event-Report-Indication]

*[Charging-Rule-Remove]

*[Charging-Rule-Install]

[Default-EPS-Bearer-QoS]

*[QoS-Information]

[Default-QoS-Information]

[Revalidation-Time]

*[Usage-Monitoring-Information]

[PCSCF-Restoration-Indication]

*[Proxy-Info]

*[Route-Record]

*[AVP]

The RAA command is sent by the enforcement point to the PCRF in response to the RAR.

<RA-Answer>::= <Diameter Header: 258, PXY >

<Session-ID>

{Origin-Host}

{Origin-Realm}

[Result-Code]

[Experimental-Result]

[Origin-State-Id]

[OC-Supported-Features]

[OC-OLR]

[IP-CAN-Type]

[RAT-Type]

[AN-Trusted]

0*2[AN-GW-Address]

[3GPP-SGSN-MCC-MNC]

[3GPP-SGSN-Address]

[3GPP-SGSN-IPv6-Address]

[RAI]

[3GPP-User-Location-Info]

[User-Location-Info-Time]

[NetLoc-Access-Support]

[User-CSG-Information]

[3GPP-MS-TimeZone]

[Default-QoS-Information]

*[Charging-Rule-Report]

[Error-Message]

[Error-Reporting-Host]

*[Failed-AVP]

*[Proxy-Info]

*[AVP]

TDF-Session-Request/Answer

The TDF-Session-Request (TSR) is sent by the PCRF to the TDF when establishing a session between the two functions. It is also used to provision rule sin the TDF once a session has been established.

<TSR> ::= <Diameter Header: 8388637, REQ, PXY>

<Session-ID>

{Vendor-Specific-Application-ID}

{Origin-Host}

{Origin-Realm}

{Destination-Realm}

[Destination-Host]

[Origin-State-ID]

[OC-Supported-Features]

*[Subscription-ID]

*[Supported-Features]

[Framed-IP-Address]

[Framed-IPv6-Prefix]

[IP-CAN-Type]

[RAT-Type]

[AN-Trusted]

[User-Equipment-Info]

[QoS-Information]

0*2[AN-GW-Address]

[3GPP-SGSN-Address]

[3GPP-SGSN-IPv6-Address]

[3GPP-GGSN-Address]

[3GPP-GGSN-IPv6-Address]

[3GPP-Selection-Mode]

[Dynamic-Address-Flag]

[Dynamic-Address-Flag-Extension]

[PDN-Connection-Charging-ID]

[3GPP-SGSN-MCC-MNC]

[RAI]

[3GPP-User-Location-Info]

[Fixed-User-Location-Info]

[User-CSG-Information]

[TWAN-Identifier]

[3GPP-MS-TimeZone]

[3GPP-Charging-Characteristics]

[Called-Station-ID]

[Charging-Information]

[Online]

[Offline]

*[ADC-Rule-Install]

[Revalidation-Time]

*[Usage-Monitoring-Information]

*[CSG-Information-Reporting]

*[Event-Trigger]

[Presence-Reporting-Area-Information]

[Logical-Access-ID]

[Physical-Access-ID]

[3GPP2-BSID]

*[Proxy-Info]

*[Route-Record]

*[AVP]

The TDF-Session-Answer (TSA) command is sent by the TDF to the PCRF when responding to the TSR.

<TSA> ::= <Diameter Header: 8388637, PXY>

<Session-ID>

{Vendor-Specific-Application-ID}

{Origin-Host}

{Origin-Realm}

[Result-Code]

[Experimental-Result]

[Origin-State-ID]

[OC-Supported-Features]

[OC-OLR]

*[Supported-Features]

*[ADC-Rule-Report]

[Event-Report-Indication]

[Error-Message]

[Error-Reporting-Host]

*[Failed-AVP]

*[Proxy-Info]

*[Route-Record]

*[AVP]

Rx AVPs

There are a number of AVPs defined specifically for the Rx interface. These AVPs are only used on the Rx, and have not been defined as common AVPs used on other interfaces. For that reason, I have defined them in this section separately. Table 7.1 lists all the AVPs used on Rx in numerical order, but the AVPs are defined in alphabetical order for ease of searching.

Abort-Cause

This is an enumerated AVP used to indicate the reason for disconnecting the bearer channel.

0 = Bearer released. Indicates the network has released this bearer channel

1 = Insufficient server resources. Indicates the server is congested and must abort the session

2 = Insufficient bearer resources. Indicates the transport gateway such as the SGSN or PDN Gateway has insufficient resources

3 = PS-to-CS handover. Indicates the bearer was released because of a handover procedure from the packet services to the circuit switched network

4 = Sponsored data connectivity disallowed. This is sent in an ASR command to indicate the PCRF is terminating a CSCF session because the operator does not support sponsored data connections when the user is roaming, or some other similar operator policy

Access-Network-Charging-Address

This AVP contains the address of the network element that is responsible for charging. This could be the address of the SGSN, or the PDN Gateway, for example.

This should not be sent through an interconnect to another network.

Access-Network-Charging-Identifier

This is a grouped AVP sent from the PCRF to the CSCF. The CSCF uses this information for correlating charging information with the IMS charging. The Access-Network-Charging-Identifier contains the GPRS charging identifier and could also provide the IP flows associated with the charging identifier.

AVP Name	Code	Value Type
Abort-Cause	500	Enumerated
Access-Network-Charging-Address	501	Address
Access-Network-Charging-Identifier	502	Grouped
Access-Network-Charging-Identifier-Value	503	OctetString
AF-Application-Identifier	504	OctetString
AF-Charging-Identifier	505	OctetString
Flow-Description	507	IPFilterRule
Flow-Number	509	Unsigned32
Flows	510	Grouped
Flow-Status	511	Enumerated
Flow-Usage	512	Enumerated
Specific-Action	513	Enumerated
Max-Requested-Bandwidth-DL	515	Unsigned32
Max-Requested-Bandwidth-UL	516	Unsigned32
Media-Component-Description	517	Grouped
Media-Component-Number	518	Unsigned32
Media-Sub-Component	519	Grouped
Media-Type	520	Enumerated
RR-Bandwidth	521	Unsigned32
RS-Bandwidth	522	Unsigned32
SIP-Forking-Indication	523	Enumerated
Codec-Data	524	OctetString
Service-URN	525	OctetString
Acceptable-Service-Info	526	Grouped
Service-Info-Status	527	Enumerated
MPS-Identifier	528	OctetString
AF-Signaling-Protocol	529	Enumerated
Sponsored-Connectivity-Data	530	Grouped
Sponsor-Identity	531	UTF8String
Application-Service-Provider-Identity	532	UTF8String
Rx-Request-Type	533	Enumerated
Min-Requested-Bandwidth-DL	534	Unsigned32
Min-Requested-Bandwidth-UL	535	Unsigned32
Required-Access-Info	536	Enumerated
IP-Domain-Id	537	OctetString
GCS-Identifier	538	OctetString
Sharing-Key-DL	539	Unsigned32
Sharing-Key-UL	540	Unsigned32
Retry-Interval	541	Unsigned32

TABLE 7.1 AVPs Used on the Rx Interface

If no IP flows are identified in the Flows AVP, then the charging identifier applies to all IP flows for the session.

<Access-Network-Charging-Identifier>::=<AVP Header; 502>

{Access-Network-Charging-Identifier-Value}

*[Flows]

Access-Network-Charging-Identifier-Value
Contains the charging identifier used by the CSCF to correlate charging records from the packet core with charging records from the IMS.

Acceptable-Service-Info
This AVP is used to request or communicate the maximum amount of bandwidth for an entire session (and all of its individual IP flows).

<Acceptable-Service-Info>::=<AVP Header: 256>

*[Media-Component-Description]

[Max-Requested-Bandwidth-DL]

[Max-Requested-Bandwidth-UL]

*[AVP]

AF-Application-Identifier
This is sent in the AAR either by itself, or as part of the Media-Component-Description AVP to describe the service being provided by the application function. It is used by the PCRF for determining the QoS it should provision in the enforcement point for the session. If this AVP appears both in the AAR and as part of the Media-Component-Description AVP, the value in the Media-Component-Description AVP shall take precedence.

AF-Charging-Identifier
This is used in the AAR for correlating charging records from the IMS with the charging records in the packet core. The CSCF sends this identifier to the PCRF so it can correlate between the charging records from both domains.

AF-Signaling-Protocol
This AVP indicates the protocol between the mobile device and the CSCF. If this is a voice call, in an LTE network, then the value should always be SIP (for VoLTE). The values are enumerated.

0 = No information

1 = SIP

Application-Service-Provider-Identity
This is used with the sponsored data connectivity AVPs to identify the application service provider for a sponsored data session.

Codec-Data

This AVP contains information about the codecs being used in the IMS for the voice call. It consists of multiple lines in ASCII. The first line will start with the word uplink or downlink, indicating how the information was received. Uplink indicates the codec information was received from the mobile device and sent to the network in the form of the Session Description Protocol (SDP). Downlink indicates the SDP information was received from the network and sent to the mobile device.

The second line starts with offer, answer, or description. A new-line character then follows each of these words. If the line starts with offer, it indicates the SDP is from an SDP line offer according to RFC 3264. If the line starts with answer, the SDP lines are from an answer. Description indicates the SDP lines came from a session description where the offer-answer mechanism defined in RFC 3264 is not being supported.

The remainder of this AVP will contain the contents of the SDP in ASCII, separated by new-line characters. The first line is "m," followed by any available "a" and "b" lines related to the "m" line. Review RFC 3264 for more information on this and the meaning of this content, which is related to the SIP protocol.

Flow-Description

The Flow-Description contains an IP filter rule, for the specified IP flow. The format is defined by IETF as an octet string, containing the following information:

- Direction as either in or out (in = uplink, out = downlink)
- Source and destination IP address
- Protocol (such as TCP)
- Source and destination port

The above information is used for filtering of packets. An action must also be included, but the only action supported in 3GPP networks is "permit." One instance of the AVP is used for each IP flow being described.

Flow-Number

This AVP contains the ordinal number for the IP flows being described in the Flow-Description AVP.

Flows

The Flows AVP contains the identifier for each of the IP flows being described in the above AVPs. If a subscriber has succeeded their quota, the Final-Unit-Action AVP describes the action to be taken for the IP flows associated with the subscriber session. This is a grouped AVP.

<Flows>::=<AVP Header: 510>

{Media-Component-Number}

*[Flow-Number]

[Final-Unit-Action]

Flow-Status
The enumerated values in this AVP indicate the status of the IP flows.

  0 = Enabled uplink

  1 = Enabled downlink

  2 = Enabled

  3 = Disabled

  4 = Removed

Flow-Usage
This AVP is used to indicate what an IP flow is being used for, and contains enumerated values. The default value is 0 (no information).

  0 = No information

  1 = RTCP

  2 = AF signaling

GCS-Identifier
This AVP is used to indicate an IMS session is related to a Group Communication Session (GCS). This is used for video broadcasting, for example, where eMBMS (LTE Broadcast) is used. The PCRF uses this information for assigning QoS and priority to each of the IP flows.

IP-Domain-Id
There are cases when a network may have multiple IP domains, and the mobile device has received an IP address in another domain (on the other side of a NAT, for example). The PCRF needs to know which domain the IP address is from when implementing session binding.

Max-Requested-Bandwidth-DL
This AVP is used to indicate the maximum bandwidth being requested for the downlink IP flow, including all protocol overhead. It is used in the AAR command to make the request. When used in the AAA, the PCRF is indicating the maximum bandwidth it will support. The value is expressed in bits per second.

Max-Requested-Bandwidth-UL
This AVP is used to indicate the maximum bandwidth being requested for the uplink IP flow, including all protocol overhead. It is used in the AAR command to make the request. When used in the AAA, the PCRF is indicating the maximum bandwidth it will support. The value is expressed in bits per second.

Media-Component-Description
This AVP is used to send information about a media session and could contain service information from a client in the mobile device. The PCRF uses this information for making determinations about the QoS to be applied to the session, and authorization. The AVP can only describe one IP flow. If there are additional IP flows, the AVP is repeated for each one.

For priority, the Reservation-Priority AVP is included to indicate the relative importance of an IP flow over other IP flows. The PCRF uses this AVP for applying priority-based admission controls. The default (if this AVP is missing) is 0, or no priority.

<Media-Component-Description>::=<AVP Header: 517>

{Media-Component-Number}

[Media-Sub-Component]

[AF-Application-Identifier]

[Media-Type]

[Max-Requested-Bandwidth-UL]

[Max-Requested-Bandwidth-DL]

[Min-Requested-Bandwidth-UL]

Min-Requested-Bandwidth-DL]

[Flow-Status]

[Reservation-Priority]

[RS-Bandwidth]

[RR-Bandwidth]

*[Codec-Data]

[Sharing-Key-DL]

[Sharing-Key-UL]

*[AVP]

Media-Component-Number
This AVP provides the order for the media component. The order is assigned based on an algorithm defined in 3GPP TS 29.214, Annex B.

Media-Sub-Component
This AVP contains the filters and bitrate for each of the IP flows in a media session. Each IP flow will have its own Media-Sub-Component. If there is conflicting values in the Media-Component-Description, this AVP takes precedence.

<Media-Sub-Component>::=<AVP Header: 519>

{Flow-Number}

0*2 [Flow-Description]

[Flow-Status]

[Flow-Usage]

[Max-Requested-Bandwidth-UL]

[Max-Requested-Bandwidth-DL]

[AF-Signaling-Protocol]

[TOS-Traffic-Class]

*[AVP]

Media-Type

The Media-Type identifies the media type for a session. The values align with the SDP used in SIP signaling, where they are derived.

0 = Audio

1 = Video

2 = Data

3 = Application

4 = Control

5 = Text

6 = Message

Min-Requested-Bandwidth-DL

This AVP indicates the minimum requested bandwidth for a session, in the downlink IP flow. If used in the AAR command, it indicates the requested bandwidth. When used in the AAA command, it indicates the authorized minimum bandwidth.

Min-Requested-Bandwidth-UL

This AVP indicates the minimum requested bandwidth for a session, in the uplink IP flow. If used in the AAR command, it indicates the requested bandwidth. When used in the AAA command, it indicates the authorized minimum bandwidth.

MPS-Identifier

This is sent by the call session control function (CSCF) to indicate the session is for Multimedia Priority Service handling. It contains the national variant for the service name. In the United States, for example, NGN GETS would be a possible value. The PCRF uses this for managing priority of sessions.

Required-Access-Info

This is used by the CSCF to notify the PCRF it needs specific access information. If user location is required, the PCRF shall send this information if available using the 3GPP-User-Location-Info, the 3GPP-SGSN-MCC-MNC AVP, the TWAN-Identifier, and the User-Location-Info-Time AVP, or the 3GPP-MS-TimeZone AVP. Not all of these AVPs have to be used but if any of them are available, the PCRF will send them to the CSCF. The values are enumerated.

0 = User location is to be reported

1 = Mobile device time zone is to be reported

Retry-Interval

When the PCRF rejects service information, it will send this AVP to the CSCF to let it know when it can retry sending the service information. It is expressed in seconds, and indicates how long the CSCF should wait before sending again.

RR-Bandwidth

The RR-Bandwidth indicates the maximum amount of bandwidth that is required for RTP Control Protocol (RTCP) receiver reports within a session component.

RS-Bandwidth

The RR-Bandwidth indicates the maximum amount of bandwidth that is required for RTCP sender reports within a session component.

Rx-Request-Type

This identifies the reason for sending an AAR command. It is enumerated.

0 = Initial request

1 = Update request

2 = PCRF restoration

Service-Info-Status

This AVP is used to indicate if service information provided by the CSCF is final, or if additional negotiation is required for the service. The PCRF can then either provision the enforcement point based on the final service information, or wait until the final status is provided. The values are enumerated.

0 = Final service information

1 = Preliminary service information

Service-URN

The Service-URN is used to indicate an emergency call or emergency session. It is sent to the PCRF for assigning priority and special consideration to the session. The values are expressed as an octet-string.

Values may include

"SOS"

"SOS.fire"

"SOS.police"

"SOS.ambulance"

Sharing-Key-DL

When media components for a subscriber can share resources, this AVP is used to identify those components in the downlink direction. The Sharing-Key-DL will be the same value for each of the shared components.

Sharing-Key-UL

When media components for a subscriber can share resources, this AVP is used to identify those components in the uplink direction. The Sharing-Key-UL will be the same value for each of the shared components.

SIP-Forking-Indication

This AVP indicates if SIP forking is being used for a session, and therefore there are several SIP dialogs related to one Diameter session. The values are enumerated.

0 = Single dialog

1 = Several dialogs

Specific-Action

This AVP is used in two different ways. When included in the RAR command, it indicates the type of action to be taken by the CSCF. When used in an AAR command, the CSCF is requesting specific actions such as notification of lost bearer channels.

0 = Not defined

1 = Charging correlation exchange

2 = Indication of loss of bearer

3 = Indication of recovery of bearer

4 = Indication of release of bearer

5 = Not defined

6 = IP-can change

7 = Indication of out of credit

8 = Indication of successful resources allocation

9 = Indication of failed resources allocation

10 = Indication of limited PCC deployment

11 = Usage report

12 = Access network information report

13 = Access network information report

14 = Indication of access network information reporting failure

Some of these values can be interpreted two ways, depending on which command they appear in. Below are the definitions for each of the values.

Charging Correlation Exchange   When used in the RAR command, it is used to report the access network charging identifier to the CSCF. When used in the AAR command, the CSCF is requesting the access network charging identifier for each of the IP flows, when it becomes known to the PCRF.

Indication of Loss of Bearer   This is used to report the loss of a bearer channels identified in the Flows AVP to the CSCF when used in the RAR command. The CSCF sends this to the PCRF in the AAR command when it is requesting notification of a loss of bearer.

Indication of Recovery of Bearer   When used in the RAR command, this value indicates that the bearer channel was recovered. The specific IP flows are identified in the Flows AVP. When used in the AAR command, the CSCF is requesting notification when the bearer channels previously lost are restored.

Indication of Release of Bearer   When used in the RAR command, this value indicates the release of a bearer channel. The IP flows are identified in the Flows AVP. When used in the AAR, the CSCF is requesting notification when a bearer channel is released.

IP-Can Change   The PCRF sends this value in the RAR command to indicate there was a change in the IP-can or RAT type. The CSCF will send this in the AAR command when requesting notification of an IP-can or RAT change.

Indication of Out of Credit   The PCRF sends this to the CSCF to indicate a session that has run out of credit (reached its quota limit). The PCRF will also send the Final-Unit-Action AVP to indicate what action is to be taken. The Flows AVP is also included to identify the IP flows that are impacted. The CSCF sends this in the AAR command to request notification when a session runs out of credit.

Indication of Successful Resources Allocation   The PCRF sends this value in the RAR command to indicate resources have been successfully allocated, if service information has been requested by the CSCF. The CSCF will send this in the AAR command to request notification of resource allocation.

Indication of Failed Resources Allocation   The PCRF will send this value in the RAR command to indicate an attempt to allocate specific resources has failed. The IP flows impacted are identified in the Flows AVP. The CSCF sends this in the AAR command to request notification of such failures.

Indication of Limited PCC Deployment   The PCRF sends this in the RAR command to indicate limited PCC deployment. The CSCF sends this in the AAR command when requesting notification of limited PCC deployment.

Usage Report   The PCRF sends this in the RAR command to report accumulated usage either by volume or time when the specified reporting threshold has been reached for sponsored data. The CSCF provides the reporting threshold. The CSCF will send this to the PCRF when requesting usage reporting for sponsored data connections.

Access Network Information Report   The PCRF sends this in the RAR command to provide the CSCF with access network information. The CSCF will send this in the AAR command when requesting notification of this report.

Indication of Recovery from Limited PCC Deployment   The PCRF sends this in the RAR command to the CSCF to report recovery from limited PCC deployment. The CSCF will send this in the AAR command when requesting notification of such recovery.

Indication of Access Network Information Reporting Failure   The PCRF sends this in the RAR command to indicate a failure in access network information reporting. The CSCF used this in the AAR command to request notification of reporting failure.

Sponsor-Identity
The Sponsor-Identity is used to identify the sponsor for a data session when sponsored data sessions are being supported. This allows service providers to provide higher quality of service to some data sessions when subsidized by a sponsor, and without charging the subscriber.

Sponsored-Connectivity-Data
The Sponsored-Connectivity-Data identifies the sponsor for a data session, the application service provider, and if the session is to be monitored (for usage), the threshold for reporting will be included.

<Sponsored-Connectivity-Data>::=<AVP Header: 530>

[Sponsor-Identity]

[Application-Service-Provider-Identity]

[Granted-Service-Unit]

[Used-Service-Unit]

*[AVP]

In addition to the AVPs shown above, there are several AVPs defined in the base protocol that are used on the Rx and may have values specific to the Rx. For example, the Experimental-Result-Code AVP and the Supported-Features AVPs use the following values on the Rx interface.

Experimental-Result-Code

Permanent Failures INVALID_SERVICE_INFORMATION (5061)—The PCRF rejects the service information sent by the CSCF because the information is invalid or insufficient
FILTER_RESTRICTIONS (5062)—The PCRF rejects the service information because the Flow-Description AVP cannot be support

REQUESTED_SERVICE_NOT_AUTHORIZED (5063)—The PCRF rejects service information because the requested service does not align with service provider policy, or it cannot be supported in the IP-CAN

DUPLICATED_AF_SESSION (5064)—The PCRF rejects a new Rx session because it relates to a session already active.

IP-CAN_SESSION_NOT_AVAILABLE (5065)—The PCRF rejects a session when the IP flow provided cannot be associated with an existing IP-CAN connection

UNAUTHORIZED_NON_EMERGENCY_SESSION (5066)—The PCRF rejects a session because session binding associated a nonemergency IMS connection to an IP-CAN connection established to an emergency APN

UNAUTHORIZED_SPONSORED_DATA_CONNECTIVITY (5067)—The PCRF rejects a session because the PCRF can't authorize the sponsored data connection

TEMPORARY_NETWORK_FAILURE (5068)—The PCRF rejects service information because a node in the access network has a temporary failure

Supported-Features
The Supported-Features AVP is actually a reused AVP, but there are specific values to the Rx interface. Those values are identified in Table 7.2.

Gx AVPs (Table 7.3)

Access-Network-Charging-Identifier-Gx AVP
This is a grouped AVP used to identify the charging identifier to be applied to a specific rule. The PCRF can use this information for correlation of charging records sent toward the CSCF in the IMS, for example.

< Access-Network-Charging-Identifier-Gx>::=< AVP Header: 1022>

{Access-Network-Charging-Identifier-Value}

*[Charging-Rule-Base-Name]

Bit	Description
0	Support of Release 8
1	Support of Release 9
2	Support of the provisioning of the AF signaling flow
3	Support for sponsored data sessions
4	Support of Release 10
5	Support of access network information reporting
6	Support of mobile device IP address being used in IP filters used in signaling between the device and the network
7	Support for sponsored data with time-based usage control
8	Support of trusted WLAN access
9	Support of release cause code information from the access network
10	Support of group communication services
11	Support of resource sharing
12	Support of deferred transfer of service information
13	Indicates the CSCF may provide a Differentiated Services Code Point (DSCP) value describing a service flow

TABLE 7.2   Supported-Features AVP

> *[Charging-Rule-Name]
> [IP-CAN-Session-Charging-Scope]
> *[AVP]

Allocation-Retention-Priority AVP
This is a grouped AVP used to indicate

- The priority of allocation and retention
- Preemption capability and preemption vulnerability

For the SDF or the EPS default bearer channel.

> <Allocation-Retention-Priority>::=<AVP Header: 1034>
> {Priority-Level}
> [Pre-emption-Capability]
> [Pre-emption-Vulnerability]

AN-GW-Address AVP
This AVP contains the IPv4 and/or the IPv6 address of the SGSN.

AN-GW-Status AVP
This enumerated AVP is sent to the PCRF to provide the status of the SGSN.

> 0 = AN_GW_FAILED

AVP Name	AVP Code	AVP Type
Access-Network-Charging-Identifier-Gx	1022	Grouped
Allocation-Retention-Priority	1034	Grouped
AN-GW-Address	1050	Address
AN-GW-Status	2811	Enumerated
APN-Aggregate-Max-Bitrate-DL	1040	Unsigned32
APN-Aggregate-Max-Bitrate-UL	1041	Unsigned32
Application-Detection-Information	1098	Grouped
Bearer-Control-Mode	1023	Enumerated
Bearer-Identifier	1020	OctetString
Bearer-Operation	1021	Enumerated
Bearer-Usage	1000	Enumerated
Charging-Correlation-Indicator	1073	Enumerated
Charging-Rule-Base-Name	1004	UTF8String
Charging-Rule-Definition	1003	Grouped
Charging-Rule-Install	1001	Grouped
Charging-Rule-Name	1005	OctetString
Charging-Rule-Remove	1002	Grouped
Charging-Rule-Report	1018	Grouped
CoA-Information	1039	Grouped
CoA-IP-Address	1035	Address
Conditional-APN-Aggregate-Max-Bitrate	2818	Grouped
Credit-Management-Status	1082	Unsigned32
CSG-Information-Reporting	1071	Enumerated
Default-EPS-Bearer-QoS	1049	Grouped
Default-QoS-Information	2816	Grouped
Default-QoS-Name	2817	UTF8String
Event-Report-Indication	1033	Grouped
Event-Trigger	1006	Enumerated
Flow-Direction	1080	Enumerated
Flow-Information	1058	Grouped
Flow-Label	1057	OctetString
Fixed-User-Location-Info	2825	Grouped
Guaranteed-Bitrate-DL	1025	Unsigned32
Guaranteed-Bitrate-UL	1026	Unsigned32
HeNB-Local-IP-Address	2804	Address
IP-CAN-Session-Charging-Scope	2827	Enumerated
IP-CAN-Type	1027	Enumerated
Metering-Method	1007	Enumerated

TABLE 7.3 Gx AVPs in Numerical Order

AVP Name	AVP Code	AVP Type
Monitoring-Flags	2828	Unsigned32
Monitoring-Key	1066	OctetString
Mute-Notification	2809	Enumerated
Monitoring-Time	2810	Time
NetLoc-Access-Support	2824	Unsigned32
Network-Request-Support	1024	Enumerated
Offline	1008	Enumerated
Online	1009	Enumerated
Packet-Filter-Content	1059	IPFilterRule
Packet-Filter-Identifier	1060	OctetString
Packet-Filter-Information	1061	Grouped
Packet-Filter-Operation	1062	Enumerated
Packet-Filter-Usage	1072	Enumerated
PCC-Rule-Status	1019	Enumerated
PDN-Connection-ID	1065	OctetString
Precedence	1010	Unsigned32
Preemption-Capability	1047	Enumerated
Preemption-Vulnerability	1048	Enumerated
Presence-Reporting-Area-Elements-List	2820	OctetString
Presence-Reporting-Area-Identifier	2821	OctetString
Presence-Reporting-Area-Information	2822	Grouped
Presence-Reporting-Area-Status	2823	Unsigned32
Priority-Level	1046	Unsigned32
PS-to-CS-Session-Continuity	1099	Enumerated
QoS-Class-Identifier	1028	Enumerated
QoS-Information	1016	Grouped
QoS-Negotiation	1029	Enumerated
QoS-Upgrade	1030	Enumerated
RAN-NAS-Release-Cause	2819	OctetString
RAT-Type	1032	Enumerated
Redirect-Information	1085	Grouped
Redirect-Support	1086	Enumerated
Reporting-Level	1011	Enumerated
Resource-Allocation-Notification	1063	Enumerated
Revalidation-Time	1042	Time
Routing-Filter	1078	Grouped
Routing-IP-Address	1079	Address

TABLE 7.3   Gx AVPs in Numerical Order (*Continued*)

AVP Name	AVP Code	AVP Type
Routing-Rule-Definition	1076	Grouped
Routing-Rule-Identifier	1077	OctetString
Routing-Rule-Install	1081	Grouped
Routing-Rule-Remove	1075	Grouped
Rule-Activation-Time	1043	Time
Rule-Deactivation-Time	1044	Time
Rule-Failure-Code	1031	Enumerated
Security-Parameter-Index	1056	OctetString
Session-Release-Cause	1045	Enumerated
TDF-Information	1087	Grouped
TDF-Application-Identifier	1088	OctetString
TDF-Application-Instance-Identifier	2802	OctetString
TDF-Destination-Host	1089	DiameterIdentity
TDF-Destination-Realm	1090	DiameterIdentity
TDF-IP-Address	1091	Address
TFT-Filter	1012	IPFilterRule
TFT-Packet-Filter-Information	1013	Grouped
ToS-Traffic-Class	1014	OctetString
Tunnel-Header-Filter	1036	IPFilterRule
Tunnel-Header-Length	1037	Unsigned32
Tunnel-Information	1038	Grouped
UDP-Source-Port	2806	Unsigned32
UE-Local-IP-Address	2805	Address
Usage-Monitoring-Information	1067	Grouped
Usage-Monitoring-Level	1068	Enumerated
Usage-Monitoring-Report	1069	Enumerated
Usage-Monitoring-Support	1070	Enumerated
User-Location-Info-Time	2812	Time
PCSCF-Restoration-Indication	2826	Unsigned32

TABLE 7.3   Gx AVPs in Numerical Order (*Continued*)

Application-Detection-Information AVP

This grouped AVP is used to report when application traffic has started and stopped, if the PCRF is subscribed to notification via Event-Triggers APPLICATION_START or APPLICATION_STOP.

<Application-Detection-Information>::=<AVP Header: 1098>

{TDF-Application-Identifier}

[TDF-Application-Instance-Identifier]

*[Flow-Information]

*[AVP]

APN-Aggregate-Max-Bitrate-DL AVP

This AVP indicates the maximum aggregate bit rate (in bits per second) to be granted for the downlink of non-GBR bearer channels with the same APN. Can be sent by the enforcement point to indicate what has been subscribed, or by the PCRF to enforce maximum bandwidth in a rule.

APN-Aggregate-Max-Bitrate-UL AVP

This AVP indicates the maximum aggregate bit rate (in bits per second) to be granted for the uplink of non-GBR bearer channels with the same APN. Can be sent by the enforcement point to indicate what has been subscribed, or by the PCRF to enforce maximum bandwidth in a rule.

Bearer-Control-Mode AVP

This is an enumerated AVP sent from the PCRF to the enforcement point indicating the bearer control mode selected by the PCRF.

0 = UE_ONLY

1 = RESERVED

2 = UE_NW

Bearer-Identifier AVP

This AVP is used to identify the bearer channel that a rule is to be applied.

Bearer-Operation AVP

This is an enumerated AVP indicating the event at the bearer channel that should trigger a request for a rule.

0 = TERMINATION

1 = ESTABLISHMENT

2 = MODIFICATION

Bearer-Usage AVP

This AVP is used to identify how the bearer channel is being used. There are only two values provided, indicating the bearer is being used for general purposes or for IMS signaling.

Charging-Correlation-Indicator AVP

This enumerated AVP is used to request the enforcement point to provide the Access-Network-Charging-Identifier-Gx AVP that is assigned to a dynamic rule. It is included in a Charging-Rule-Install AVP.

0 = CHARGING_IDENTIFIER_REQUIRED

Charging-Rule-Install AVP

This is a grouped AVP used to activate, install or modify rules sent by the PCRF to the enforcement point. When installing a new rule, the PCRF would send this AVP with the Charging-Rule-Definition AVP to install the new rule.

If a rule is already installed at the enforcement point, the PCRF sends the Charging-Rule-Name AVP to identify the rule to be activated. It can also send the Charging-Rule-Base-Name AVP if activating a group of rules.

The Bearer-Identifier AVP is used to identify a bearer channel that the rule is to be applied to, if the rule is to be applied to an IP-CAN session.

<Charging-Rule-Install>::=<AVP Header: 1001>

*[Charging-Rule-Definition]

*[Charging-Rule-Name]

*[Charging-Rule-Base-Name]

[Bearer-Identifier]

[Monitoring-Flags]

[Rule-Activation-Time]

[Rule-Deactivation-Time]

[Resource-Allocation-Notification]

[Charging-Correlation-Indicator]

*[AVP]

Charging-Rule-Definition AVP

This grouped AVP is sent by the PCRF to define a rule to be installed or activated by the enforcement point. The Flow-Identifier indicates the specific data flow that the rule is to be applied. If the Flow-Identifier is not provided, the TDF-Application-Identifier AVP must be provided to indicate an application filter.

The Flows AVP can only be used if the AF-Charging-Identifier AVP is also provided. The AF-Signaling-Protocol AVP will be included if the rule applies to IMS signaling.

If the Reporting-Level AVP is set to the value of SPONSORED_CONNECTIVITY_ LEVEL, the Sponsor-Identity and Application-Service-Provider AVPs will also be included in the message.

<Charging-Rule-Definition>::=<AVP Header: 1003>

{Charging-Rule-Name}

[Service-Identifier]

[Rating-Group]

*[Flow-Information]

[TDF-Application-Identifier]

[Flow-Status]

[QoS-Information]

[PS-to-CS-Session-Continuity]

[Reporting-Level]

[Online]

[Offline]

[Metering-Method]

[Precedence]

[AF-Charging-Identifier]

*[Flows]

[Monitoring-Key]

[Redirect-Information]

[Mute-Notification]

[AF-Signaling-Protocol]

[Sponsor-Identity]

[Application-Service-Provider-Identity]

*[Required-Access-Info]

[Sharing-Key-DL]

[Sharing-Key-UL]

*[AVP]

Charging-Rule-Name AVP

This AVP is used to identify a rule. It is sent by the PCRF to the enforcement point when the PCRF activating or installing rules on the enforcement point.

Charging-Rule-Remove AVP

This grouped AVP is used when removing or deactivating a rule at the enforcement point. The Charging-Rule-Name AVP identifies the rule to be removed or deactivated. The Required-Access-Info AVP is only used when the CSCF in the IMS is requesting the PCRF to send network access information (such as location) for the mobile device, and the CSCF is requesting the rules to be removed.

<Charging-Rule-Remove>::=<AVP Header: 1002>

*[Charging-Rule-Name]

*[Charging-Rule-Base-Name]

*[Required-Access-Info]

*[AVP]

Charging-Rule-Report AVP

This is a grouped AVP used to report the status of a rule. This includes rules that could not be installed or activated at the enforcement point, or rules that could not be enforced. Multiple instances of this AVP can be used in a single command.

<Charging-Rule-Report>::=<AVP Header: 1018>

*[Charging-Rule-Name]

*[Charging-Rule-Base-Name]

[Bearer-Identifier]

[PCC-Rule-Status]

[Rule-Failure-Code]

[Final-Unit-Indication]

*[RAN-NAS-Release-Cause]

*[AVP]

CoA-Information AVP

This grouped AVP is sent from the enforcement point to the PCRF, and contains the care-of-address and tunnel-information related to the care-of-address.

<CoA-Information>::=<AVP Header: 1039>

{Tunnel-Information}

{CoA-IP-Address}

*[AVP]

CoA-IP-Address AVP

This AVP provides the mobile devices care-of-address providing either an IPv4 or IPv6 address.

Conditional-APN-Aggregate-Max-Bitrate AVP

This grouped AVP is used to define the APN policy information.

<Conditional-APN-Aggregate-Max-Bitrate>::=<AVP Header: 2818>

[APN-Aggregate-Max-Bitrate-UL]

[APN-Aggregate-Max-Bitrate-DL]

*[IP-CAN-Type]

*[RAT-Type]

*[AVP]

Credit-Management-Status AVP

This is a bit-masked AVP used for quota management (Table 7.4).

CSG-Information-Reporting AVP

This is an enumerated AVP used by the PCRF to instruct the enforcement point to report changes in the closed subscriber group (CSG) information to the offline charging system (OFCS). Since this reporting could generate large amounts of signaling, it is recommended that this reporting be limited to a small portion of subscribers.

0 = CHANGE_CSG_CELL

Indicates the enforcement point should report to the OFCS whenever the mobile device changes its access using a CSG cell

1 = CHANGE_CSG_SUBSCRIBED_HYBRID_CELL

Bit	Name	Description
0	Service denied	The charging system denied service because of limitations applied to the end user. Usually this happens when there is not enough credit on account.
1	Credit control does not apply	Service is granted but quota management is not required because the service is being offered at no charge, for example.
2	Authorization rejected	Service is denied by the charging system because the service is being terminated.
3	User unknown	Indicates the subscriber is unknown to the charging system
4	Rating failed	The charging system is unable to rate the session because there is not enough information provided (such as rating input, an incorrect AVP or AVP values that are not correct).
5	No Gyn session, service allowed	The Gyn session was terminated, but the OCS is allowing the service to continue.
6	No Gyn session, service not allowed	The Gyn session was terminated, and the OCS is not allowing the service to continue.

TABLE 7.4   Credit-Management-Status AVP Values

Indicates the enforcement point should report to the OFCS whenever the mobile device changes its access using a hybrid cell to which it is a subscriber.

2 = CHANGE_CSG_UNSUBSCRIBED_HYBRID_CELL

Indicates the enforcement point should report to the OFCS whenever the mobile device changes its access using a hybrid cell to which it is not a subscriber.

Default-EPS-Bearer-QoS AVP
This is a grouped AVP used to define the QoS for a default bearer channel.

<Default-EPS-Bearer-QoS>::=<AVP Header: 1049>

[QoS-Class-Identifier]

[Allocation-Retention-Priority]

*[AVP]

Default-QoS-Information AVP
This grouped AVP is used to define the default QoS information for an IP-CAN session related to fixed broadband access.

<Default-QoS-Information>::=<AVP Header: 2816>

[QoS-Class-Identifier]

[Max-Requested-Bandwidth-UL]

[Max-Requested-Bandwidth-DL]

[Default-QoS-Name]

*[AVP]

Default-QoS-Name AVP

Indicates the name for a preconfigured QoS profile configured in the enforcement point.

Event-Report-Indication AVP

This is a grouped AVP used by the PCRF to report an event from the bearer binding and event reporting function (BBERF) to the enforcement point. The enforcement point is used to provide information about the required event triggers to the PCRF. When sent by the enforcement point to the PCRF, only the Event-Trigger AVP is sent using the below possible values:

- RAI_CHANGE
- RAT_CHANGE
- USER_LOCATION_CHANGE
- UE_TIME_ZONE_CHANGE
- USER_CSG_INFORMATION_CHANGE
- USER_CSG_HYBRID_SUBSCRIBED_INFORMATION_CHANGE
- USER_CSG_HYBRID_UNSUBSCRIBED_INFORMATION_CHANGE
- TAI_CHANGE

<Event-Report-Indication>::=<AVP Header: 1033>

*[Event-Trigger]

[User-CSG-Information]

[RAT-Type]

[RAI]

[3GPP-User-Location-Info]

[Trace-Data]

[Trace-Reference]

[3GPP2-BSID]

[3GPP-MS-TimeZone]

[Routing-IP-Address]

*[AVP]

Note: Several AVPs have been omitted, as they are not applicable to the Gx interface.

Event-Trigger AVP

The event trigger is used in a few different ways. When the PCRF sends the Event-Trigger AVP to the enforcement point, it is used to indicate what events will cause the enforcement point to request rules from the PCRF to be applied to the event. When sent by the enforcement point to the PCRF, it is used by the enforcement point to indicate the event has occurred at the enforcement point.

The values are enumerated, and defined as follows:

0 = SGSN CHANGE

This trigger is used when the serving gateway (such as the SGSN) changes to another gateway. The PCRF sends this to the enforcement point when responding to a CCR command, or by sending an RAR. The enforcement point uses this value to send notification to the PCRF when the event occurs. The 3GPP-SGSN-Address or the 3GPP-SGSN-IPv6-Address AVP is used to send the new address.

1 = QOS CHANGE

This trigger is used to notify the PCRF that there was either a change in the requested QoS for a specified bearer channel, or a change in the QoS for a specific APN.

2 = RAT CHANGE

The PCRF sends this to the enforcement point in the CCA or the RAR command to indicate rules are to be requested when the radio access type changes. The enforcement point uses this AVP in the CCR command along with the RAT-Type AVP to request rules from the PCRF when the radio access type changed.

3 = TFT CHANGE

The PCRF sends this to the enforcement point in the CCA or the RAR command to indicate the enforcement point is to request rules when the traffic filter template (TFT) changes. The enforcement point will sue the CCR command when the TFT changes at the bearer. The Bearer-Identifier AVP is used to identify the bearer. The TFT filter definitions are provided in the TFT-Packet-Filter-Information AVP.

4 = PLMN CHANGE

The PCRF sends this to the enforcement point sing the CCA or the RAR command, indicating the enforcement point requests rules when the PLMN changes. The enforcement point sends this value using the CCR command when there is a change in the PLMN. The new network is identified using the 3GPP-SGSN-MCC-MNC AVP.

5 = LOSS OF BEARER

This is sent in the CCA or the RAR command by the PCRF to the enforcement point, indicating the enforcement point is to notify the PCRF when there is a loss of bearer. The enforcement point sends this in the CCR command when notifying the PCRF of a loss of bearer. The rules associated with the lost bearer are identified in the Charging-Rule-Report AVP.

The PCC-Rule-Status AVP is sent in the Charging-Rule-Report AVP with the value of INACTIVE.

6 = RECOVERY OF BEARER

This is sent by the PCRF in the CCA or RAR commands to the enforcement point when it wishes to receive notification that a bearer channel has been restored. The enforcement point sends this in the CCR command when notifying the PCRF that a bearer channel has been restored.

7 = IP-CAN_CHANGE

The PCRF sends this value in the CCA or RAR command when requesting notification of changes to an IP-CAN. The enforcement point sends this in the CCR command to request new rules, because the IP-CAN has changed, and the PCRF has indicated the enforcement point is to request new rules when there is a change.

11 = QoS_CHANGE_EXCEEDING_AUTHORIZATION

The enforcement point sends this value in a CCR command to the PCRF if there has been a change in the requested QoS beyond what is currently authorized for a bearer session. The PCRF requests this notification by sending this value in a CCA or RAR command.

12 = RAI_CHANGE

This is sent by the enforcement point when there is a change in the routing area identification (RAI) and the PCRF has previously notified the enforcement point to request new rules should there be a change in the RAI. The new RAI value is provided in the RAI AVP.

13 = USER_LOCATION_CHANGE

The PCRF sends this value in a CCA or RAR command to indicate the enforcement point is to request new rules when there is a change in a user's location. The enforcement point sends this value along with the 3GPP-User-Location-Info AVP indicating the new location for the user.

14 = NO_EVENT_TRIGGERS

This value is sent by the PCRF in the CCA or RAR command to indicate to the enforcement point notification of an event is not required, except for those events that are always provisioned in the enforcement point.

15 = OUT_OF_CREDIT

This is sent by the PCRF in the CCA or RAR command to indicate to the enforcement point it is to request new rules when credit for a rule is no longer available. The PCRF will also indicate the action to be taken by the enforcement point when credit expires (such as terminate the session). The enforcement point sends this value in the CCR when credit is exhausted for a session. The enforcement point also sends the Charging-Rule-Report AVP as well as the Final-Unit-Indication AVP.

16 = REALLOCATION_OF_CREDIT

The PCRF uses this value to indicate to the enforcement point it is to notify the PCRF of rules where credit has been reallocated after previously running out of credit. The enforcement point sends this value along with the Charging-Rule-Report AVP to indicate the specified rule has been reallocated credit.

17 = REVALIDATION_TIMEOUT

This value is used to indicate when there has been a revalidation timeout for a rule. It is sent by the PCRF to indicate notification is necessary and used by the enforcement point in the CCR command when notifying the PCRF of the event.

18 = UE_IP_ADDRESS_ALLOCATE

This is sent by the enforcement point when there is an IPv4 address allocated for a mobile device. The Framed-IP-Address AVP is also sent to provide the IP address. The PCRF does not provision this in the enforcement point; therefore, the PCRF will not send this to the enforcement point.

19 = UE_IP_ADDRESS_RELEASE

The enforcement point uses this to notify the PCRF that an IP address previously allocated for a mobile device has been released. The Framed-IP-Address AVP is also provided to indicate the IP address that was released.

20 = DEFAULT_EPS_BEARER_QOS_CHANGE

This value is used to indicate there was a change in the default EPS bearer QoS. The PCRF uses this to request notification when there is a change. The new value for the QoS is provided in the Default-EPS-Bearer-QoS AVP.

21 = AN_GW_CHANGE

This value is used by the PCRF to request notification when the serving access network node gateway (such as the SGSN) changes. It indicates to the enforcement point that a CCR should be sent requesting rules. The enforcement point will send this to the PCRF in the CCR with the address of the new serving access network node in the AN-GW-Address AVP.

22 = SUCCESSFUL_RESOURCE_ALLOCATION

Some rules provisioned by the PCRF may require notification to the PCRF when the resources have been allocated for the rule. The PCRF requests this notification using this value, and the enforcement point uses this value to inform the PCRF of resource allocation. The affected rules are indicated in the Charging-Rule-Report by setting the PCC-Rules-Status AVP to ACTIVE.

23 = RESOURCE_MODIFICATION_REQUEST

This value is sent by the enforcement point to report to the PCRF that an event occurred. The event resulted in modification to resources as defined in the Packet-Filter-Operation and Packet-Filter-Information AVPs.

24 = PGW_TRACE_CONTROL

This is sent to indicate that the command it is contained in also contains a trace activation of deactivation request for the PDN Gateway. The trace parameters for activation are included in the Trace-Data AVP. When deactivating traces, the traces to be deactivated are indicated in the Trace-Reference AVP.

25 = UE_TIME_ZONE_CHANGE

This value is used to indicate to the enforcement point that the PCRF needs to be notified when the mobile device changes time zones (because it is roaming). The enforcement point will send this to the PCRF to indicate that there was a change in the time zone for the mobile device. The new time zone is provided in the 3GPP-MS-TimeZone AVP.

26 = TAI_CHANGE

This is used to notify the PCRF when the mobile device changes tracking area identities (TAI) due to roaming. The PCRF requests the notification by sending this trigger to the enforcement point and the enforcement point sends this trigger value to notify the PCRF when there is a change to the TAI. The enforcement point provides the new value for the TAI in the 3GPP-User-Location-Info AVP.

27 = ECGI_CHANGE

This trigger value is used when PCRF is requesting notification of a change in the ECGI. The enforcement point uses this value to inform the PCRF of the change and includes the new ECGI value in the 3GPP-User-Location-Info AVP.

28 = CHARGING_CORRELATION_EXCHANGE

The PCRF uses this trigger value to request the access network charging identifier that is associated with one or more rules. The Charging-Correlation-Indicator AVP value of CHARGING_IDENTIFIER_REQUIRED is included. The enforcement point sends this trigger value when the charging identifier has been assigned.

29 = APN-AMBR_MODIFICATION_FAILURE

The enforcement point sends this trigger value to the PCRF to notify it that APN-AMBR modifications failed.

30 = USER_CSG_INFORMATION_CHANGE

This trigger is used when a subscriber enters a closed subscriber group cell. The User-CSG-Information AVP is provided along with the notification to the PCRF.

33 = USAGE_REPORT

This trigger is sent by the PCRF when requesting usage monitoring from the enforcement point. The PCRF also sends the Usage-Monitoring-Information AVP that contains the Monitoring-Key and Granted-Service-Unit AVPs. The enforcement point in turn uses this value when reporting usage.

34 = DEFAULT-EPS-BEARER-QOS_MODIFICATION_FAILURE

This trigger is only sent by the enforcement point to the PCRF to indicate a failure of the default bearer channel QoS modifications.

35 = USER_CSG_HYBRID_SUBSCRIBED_INFORMATION_CHANGE

This trigger is used to report to the PCRF that a subscriber has either entered or is leaving a closed subscriber group (CSG) hybrid cell to which the user subscribes. The User-CSG-Information AVP is provided as well.

36 = USER_CSG_ HYBRID_UNSUBSCRIBED_INFORMATION_CHANGE

This trigger is used to report to the PCRF that a subscriber has either entered or is leaving a closed subscriber group (CSG) hybrid cell to which the user does not subscribe. The User-CSG-Information AVP is provided as well.

37 = ROUTING_RULE_CHANGE

This trigger is used to notify the PCRF when there is a change in the routing rules for IP flow mobility. No subscription is necessary from the PCRF, so this is not sent by PCRF to enforcement point. The enforcement point will include the routing rule change in the Routing-Rule-Definition AVP when sending to the PCRF.

39 = APPLICATION_START

When the enforcement point is monitoring application usage, it reports when application traffic starts. The Application-Detection-Information AVP identifies the application traffic that has started.

40 = APPLICATION_STOP

This trigger is used to indicate application traffic has stopped, when being reported by the enforcement point. The PCRF sends this trigger when requesting notification of application traffic start and stop.

42 = CS_TO_PS_HANDOVER

This is the trigger used to notify the PCRF when there is a handover from the circuit-switched network to the packet-switched network.

43 = UE_LOCAL_IP_ADDRESS_CHANGE

The enforcement point will use this trigger to notify the PCRF when the local IP address or the UDP source port number, or both, has changed for a mobile device. The UE-Local-IP-Address and/or the UDP-Source-Port AVPs are also provided to identify the new addresses. The PCRF does not request this from the enforcement point so this trigger is not used in a CCA or RAR.

44 = H(E)NB_LOCAL_IP_ADDRESS_CHANGE

This is used by the enforcement point to notify the PCRF when the H(e)NB local IP address or the UDP source port number (or both) has changed. The

HeNB-Local-IP-Address AVP and/or the UDP-Source-Port AVP are also sent to communicate the new address.

45 = ACCESS_NETWORK_INFO_REPORT

When network location information is needed to support policies, the PCRF may request the enforcement point to provide this information using access network information procedures. This trigger is used by the PCRF to request notification and the location information and it is used by the enforcement point to provide this information to the PCRF.

46 = CREDIT_MANAGEMENT_SESSION_FAILURE

The enforcement point uses this trigger to notify the PCRF when there was either a transient or a permanent failure in the OCS. The enforcement point will provide the Charging-Rule-Report AVP containing the PCC-Rules-Status value ACTIVE and the Rule-Failure-Code AVP providing the failure cause code as provided by the OCS, when the failure applies only to select rules and not all rules.

47 = DEFAULT_QOS_CHANGE

The enforcement point uses this trigger to indicate a change in the default QoS to the PCRF. The enforcement point will include the Default-QoS-Information AVP with the new value. The PCRF sends this trigger to the enforcement point when requesting this notification.

48 = CHANGE_OF_UE_PRESENCE_IN_PRESENCE_REPORTING_AREA_REPORT

This trigger indicates that the mobile device is within or outside of the presence reporting area. The presence reporting area is identified in the Presence-Reporting-Area-Information AVP when the trigger is provisioned in the enforcement point by the PCRF.

Fixed-User-Location-Info AVP
This grouped AVP provides the location of user equipment in a fixed broadband network.

<Fixed-User-Location-Info>::=<AVP Header: 2825>

[SSID]

[BSSID]

[Logical-Access-ID]

[Physical-Access-ID]

*[AVP]

Flow-Direction AVP
This enumerated AVP is used to indicate the direction that a filter is applied.

0 = UNSPECIFIED

1 = DOWNLINK

2 = UPLINK

3 = BIDIRECTIONAL

Flow-Information AVP

This is a grouped AVP sent by the PCRF containing information about a single IP flow filter. If the Flow-Description, ToS-Traffic-Class, Security-Parameter-Index or Flow-Label AVPs are present, the Flow-Direction AVP must also be included.

< Flow-Information>::=< AVP Header: 1058>

[Flow-Description]

[Packet-Filter-Identifier]

[Packet-Filter-Usage]

[ToS-Traffic-Class]

[Security-Parameter-Index]

[Flow-Label]

[Flow-Direction]

*[AVP]

Flow-Label AVP

This AVP contains the IPv6 flow label header.

Guaranteed-Bitrate-DL AVP

This AVP is used to indicate the guaranteed bitrate (GBR) for a downlink data flow. The bandwidth specified includes all protocol overhead from IP and transport layers, as well as payload.

Guaranteed-Bitrate-UL AVP

This AVP is used to indicate the guaranteed bitrate (GBR) for an uplink data flow. The bandwidth specified includes all protocol overhead from IP and transport layers, as well as payload.

HeNB-Local-IP-Address AVP

This AVP provides the IP address for the home eNodeB.

IP-CAN-Session-Charging-Scope AVP

This is an enumerated AVP that simply indicates that the access network charging identifier is to be applied to an entire IP-CAN session.

0 = IP-CAN_SESSION_SCOPE

IP-CAN-Type AVP

This is an enumerated AVP indicating the type of connection being used by the subscriber to access the network.

0 = 3GPP-GPRS

1 = DOCSIS

2 = xDSL

3 = WiMAX

4 = 3GPP2

5 = 3GPP-EPS

6 = Non-3GPP-EPS

7 = FBA

Metering-Method AVP

This is an enumerated AVP providing the method to be applied for metering of a session in offline charging. It can also be used by the enforcement point for requesting units used in online charging when decentralized unit determination is being used. The values are:

0 = DURATION

Indicates the duration of the data flow traffic is metered.

1 = VOLUME

Indicates the volume of the data flow traffic is metered.

2 = DURATION_VOLUME

Indicates that both duration and volume are to be used for metering.

3 = EVENT

Indicates that event-based charging is to be applied to the data flow traffic.

Monitoring-Flags AVP

This AVP is bit-masked and only has one value. When this first bit (bit 0) is set to the value of 1, it indicates that the data flow is not going to be included in the volume or time measurements at the IP-CAN session level.

Monitoring-Key AVP

This AVP is used to identify usage monitoring instances. Each monitor is given an identifier for referencing by the PCRF and the enforcement point.

Monitoring-Time AVP

Used by the PCRF to specify what time the enforcement point is to reapply a threshold for a rule.

Mute-Notification AVP

This is an enumerated AVP that is used to mute notification of an application's start and stop time.

0 = MUTE_REQUIRED

NetLoc-Access-Support AVP

This AVP is used to indicate if the access network supports network location procedures.

0 = NETLOC_ACCESS_NOT_SUPPORTED

Network-Request-Support AVP

This is an enumerated AVP used to indicate that the mobile device and the network support network initiated procedures.

0 = NETWORK_REQUEST NOT SUPPORTED

1 = NETWORK_REQUEST SUPPORTED

Offline AVP

This is an enumerated AVP used within the Charging-Rule-Definition AVP to define whether the offline charging interface at the enforcement point is to be enabled for a rule.

0 = DISABLE_OFFLINE

Indicates the offline charging interface is to be disabled.

1 = ENABLE_OFFLINE

Indicates the offline charging interface is to be enabled.

Online AVP

This enumerated AVP indicates whether or not the online charging interface at the enforcement point is to be enabled for a specific rule. It is sent along with the Charging-Rule-Install AVP as a rule is being provisioned, or it can be sent in the CCR at the command level (not as part of a grouped AVP). If it is sent at the command level, it indicates that whatever the default online charging method preconfigured in the enforcement point should be applied. The following values are defined:

0 = DISABLE_ONLINE

Indicates that online charging is to be disabled for the associated rule.

1 = ENABLE_ONLINE

Indicates that online charging is to be enabled for the associated rule.

Packet-Filter-Content AVP

This AVP contains the packet filter contents requested by the mobile device. This content is needed by the PCRF to create rules. The information shall include

- Action ("permit")
- Direction (out)
- Protocol (IP)
- Source IP address

- Source and/or destination port
- Destination IP address

Packet-Filter-Identifier AVP

This AVP indicates the identity of the packet filter. This is assigned by the PCRF for rules created when a mobile device initiates a connection.

Packet-Filter-Information AVP

This is a grouped AVP containing the information from a packet filter. It is sent by the enforcement point to the PCRF.

< Packet-Filter-Information>::=< AVP Header: 1061>

[Packet-Filter-Identifier]

[Precedence]

[Packet-Filter-Content]

[ToS-Traffic-Class]

[Security-Parameter-Index]

[Flow-Label]

[Flow-Direction]

*[AVP]

Packet-Filter-Operation AVP

This enumerated AVP indicates a mobile device has initiated a connection requiring resources that require rules to be assigned.

0 = DELETION

1 = ADDITION

2 = MODIFICATION

Packet-Filter-Usage AVP

This enumerated AVP indicates whether or not the mobile device is to be provisioned with traffic mapping information.

1 = SEND_TO_UE

PCC-Rule-Status AVP

This is an enumerated AVP describing the status of a rule or a group of rules.

0 = ACTIVE

1 = INACTIVE

2 = TEMPORARILY_INACTIVE

PCSCF-Restoration-Indication AVP
This AVP is used to request P-CSCF restoration.

0 = PCSCF_RESTORATION

PDN-Connection-ID AVP
This AVP is used to provide the PDN connection identifier.

Precedence AVP
The precedence AVP can be used in two ways. When it is included in the Charging-Rule-Definition AVP, it determines what order data flow templates (which contain data flow filters) are applied to traffic at the enforcement point.

For rules using an application detection filter, precedence determines what rule is to be applied to the detected application traffic for QoS, charging control, reporting start and stop of the traffic flow, and usage monitoring. The lowest value represents the highest precedence.

When it is included in the TFT-Packet-Filter-Information AVP, it indicates the evaluation order for traffic received from the mobile device.

When it is used in the Routing-Rule-Definition AVP, it indicates the evaluation order of routing filters in the IP flow routing rules.

Preemption-Capability AVP
This is an enumerated AVP indicating if the default bearer can be granted resources already assigned to another bearer channel with a lower priority level.

0 = PRE-EMPTION_CAPABILITY_ENABLED

1 = PRE-EMPTION_CAPABILITY_DISABLED

Preemption-Vulnerability AVP
This is an enumerated AVP indicating whether or not a data flow can be preempted by another data flow with a higher priority.

0 = PRE-EMPTION_VULNERABILITY_ENABLED

1 = PRE-EMPTION_VULNERABILITY_DISABLED

Presence-Reporting-Area-Elements-List AVP
This AVP is used in presence services to allow a mobile device reporting back to the network its presence status. The values provided in this AVP are found in the presence reporting area action as stated in TS 29.274 (Table 7.5). The parameters identified are

Presence-Reporting-Area-Identifier AVP
Presence services allow the network to report the availability of a mobile device based on the status reported by the device. The presence reporting area is identified in this AVP by a unique identifier.

Octet		
9	Number of TAI	Number of RAI
10	Number of macro eNodeB	
11	Number of home eNodeB	
12	Number of ECGI	
13	Number of SAI	
14	Number of CGI	
15 to k	TAIs (1–15)	
k+1 to m	Macro eNB IDs (1–63)	
m+1 to p	Home eNB IDs (1–63)	
p+1 to q	ECGIs (1–63)	
q+1 to r	RAIs (1–15)	
r+1 to s	SAIs (1–63)	
s+1 to t	CGIs (1–63)	

TABLE 7.5   AVP Values for Presence-Reporting-Area-Elements-List AVP

Presence-Reporting-Area-Information AVP
This grouped AVP is used in presence services to provide status of a mobile device, as well as information about the reporting area the mobile device is reporting from, and other presence service parameters.

<Presence-Reporting-Area-Information>::=<AVP Header: 2822>

[Presence-Reporting-Area-Identifier]

[Presence-Reporting-Area-Status]

[Presence-Reporting-Area-Elements-List]

*[AVP]

Presence-Reporting-Area-Status AVP
This AVP is used to indicate if the mobile device is in or out of the presence reporting area.

0 = In presence reporting area

1 = Out of presence reporting area

Priority-Level AVP
The priority level value can be from 1 to 15, with one being the highest priority. The priorities are assigned to rules, and determine whether bearer establishment or modification can be supported based on resource limitations. Values 1 to 8 should be reserved for services authorized within an operator domain, while priorities 9 to 15 should be reserved for roamers.

PS-to-CS-Session-Continuity AVP

This enumerated AVP indicates if a data flow is a candidate for packet services to circuit-switched continuity.

0 = VIDEO_PS2CS_CONT_CANDIDATE

QoS-Class-Identifier AVP

This is an enumerated AVP that identifies the QCI to be applied to an IP-CAN or data flow. The QCI is standardized and the values shown in Table 7.6.

Value	QCI Value	Resource	Priority	Description
0	Reserved			
1	QCI_1	GBR	2	Conversational voice
2	QCI_2		4	Conversational video (live streaming)
3	QCI_3		3	Real time gaming
4	QCI_4		5	Nonconversational video (buffered streaming)
5	QCI_5	Non GBR	1	IMS signaling
6	QCI_6		6	Video (buffered streaming), TCP-based (e.g., WWW, e-mail, chat, FTP, P2P file sharing, progressive video, etc.)
7	QCI_7		7	Voice, video (live streaming), interactive gaming
8	QCI_8		8	Video (buffered streaming), TCP-based (e.g., WWW, e-mail, chat, FTP, P2P file sharing, progressive video, etc.)
9	QCI_9		9	Video (buffered streaming), TCP-based (e.g., WWW, e-mail, chat, FTP, P2P file sharing, progressive video, etc.)
10–64	Spare			
65	QCI_65	GBR	0.7	Mission Critical Push To Talk voice (typically for emergency services)
66	QCI_66		2	Mission Critical Push To Talk voice (typically for emergency services)
67–68	Spare			
69	QCI_69	Non GBR	0.5	Mission Critical Push To Talk voice (typically for emergency services)
70	QCI_70		5.5	Mission Critical Push To Talk voice (typically for emergency services)
71–127	Spare			
128–254				Operator specific

TABLE 7.6 QCI Values for LTE Networks

QoS-Information AVP

This is a grouped AVP defining the QoS to be applied to resources requested by the mobile device. Resources can be an IP-CAN bearer channel, a rule, QCI, or APN. The enforcement point sends this to the PCRF when requesting QoS information for a resource, and the PCRF responds to the request by sending the QoS information in this AVP.

<QoS-Information>::=<AVP Header: 1016>

[QoS-Class-Identifier]

[Max-Requested-Bandwidth-UL]

[Max-Requested-Bandwidth-DL]

[Guaranteed-Bitrate-UL]

[Guaranteed-Bitrate-DL]

[Bearer-Identifier]

[Allocation-Retention-Priority]

[APN-Aggregate-Max-Bitrate-UL]

[APN-Aggregate-Max-Bitrate-DL]

*[Conditional-APN-Aggregate-Max-Bitrate]

*[AVP]

QoS-Negotiation AVP

This is an enumerated AVP indicating if the PCRF is allowed to negotiate QoS by responding to a request with a QoS value other than what was provided in the request.

0 = NO_QoS_NEGOTIATION

1 = QoS_NEGOTIATION_SUPPORTED

QoS-Upgrade AVP

This is an enumerated AVP indicating whether or not the SGSN supports the GGSN upgrading the QoS when creating or updating a PDP context.

0 = QoS_UPGRADE_NOT_SUPPORTED

1 = QoS_UPGRADE_SUPPORTED

RAN-NAS-Release-Cause AVP

This AVP is used to send the cause code indicating the reason a 3GPP EPS session was released. Also applies to non-3GPP networks.

RAT-Type AVP

This enumerated AVP identifies the radio access technology used by the mobile device to access the network.

0 = WLAN

1 = VIRTUAL

2 – 999 = Generic technologies that are not IP-CAN specific

1000 = UTRAN

1001 = GERAN

1002 = GAN

1003 = HSPA_EVOLUTION

1004 = EUTRAN

1005 – 1999 = 3GPP specific radio technology

2000 = CDMA2000_1X

2001 = HRPD

2002 = UMB

2003 = EHRPD

2004 – 2999 = 3GPP2 specific radio technology

Redirect-Information AVP

This grouped AVP provides the address where detected application traffic is to be redirected. It is sent by the PCRF in the Charging-Rule-Definition AVP.

<Redirect-Information>::=<AVP Header: 1085>

[Redirect-Support]

[Redirect-Address-Type]

[Redirect-Server-Address]

*[AVP]

Redirect-Support AVP

This enumerated AVP indicates if redirection is enabled or disabled.

0 = REDIRECTION_DISABLED

1 = REDIRECTION_ENABLED

Reporting-Level AVP

This is an enumerated AVP defining the level of reporting of usage for a rule.

0 = SERVICE_IDENTIFIER_LEVEL

Indicates that usage reporting is to include the service identifier and the rating group, when the Service-Identifier and the Rating-Group AVPs have been included in the Charging-Rule-Definition.

1 = RATING_GROUP_LEVEL

Indicates usage reporting on the rating group level is to be used, when the Rating-Group AVP has been included in the Charging-Rule-Definition AVP.

2 = SPONSORED_CONNECTIVITY_LEVEL

Indicates that usage shall be reported on the sponsor identity and rating group when the Sponsor-Identity, Application-Service-Provider-Identity, and Rating-Group AVPs are included in the Charging-Rule-Definition AVP. This is only applicable for offline charging.

Resource-Allocation-Notification AVP

This enumerated AVP indicates whether or not notification is to be provided to the PCRF when resources are allocated for rules. The rules are identified in the Charging-Rule-Install AVP.

0 = ENABLE_NOTIFICATION

Revalidation-Time AVP

This AVP is used by the PCRF to indicate when the enforcement point will need to re-request rules from the PCRF. The time is NTP and only applied when the Event-Trigger REVALIDATION_TIMEOUT has been sent in the same command (or has already been provisioned).

Routing-Filter AVP

This grouped AVP is used by the enforcement point to send information to the PCRF for a routing filter.

<Routing-Filter>::=<AVP Header: 1078>

{Flow-Description}

{Flow-Direction}

[ToS-Traffic-Class]

[Security-Parameter-Index]

[Flow-Label]

*[AVP]

Routing-IP-Address AVP

This AVP contains the IPv4 or IPv6 home address or the care-of-address for a mobile node.

Routing-Rule-Definition AVP

This is a grouped AVP used to define the IP flow routing rule being sent by the enforcement point to the PCRF.

<Routing-Rule-Definition>::=<AVP Header: 1076>

{Routing-Rule-Identifier}

*[Routing-Filter]

[Precedence]

[Routing-IP-Address]

*[AVP]

Routing-Rule-Identifier AVP

This is used to provide a unique identifier for each IP flow routing rule. The enforcement point assigns this identifier and uses it when communicating to the PCRF regarding actions to be taken with this rule.

Routing-Rule-Install AVP

This grouped AVP is used to either install or modify IP flow routing rules.

> <Routing-Rule-Install>::=<AVP Header: 1081>
>
> *[Routing-Rule-Definition]
>
> *[AVP]

Routing-Rule-Remove AVP

This grouped AVP is used for removing IP flow routing rules from the PCRF. The rule to be removed is identified in the Routing-Rule-Identifier AVP.

> <Routing-Rule-Remove>::=<AVP Header: 1075>
>
> *[Routing-Rule-Identifier]
>
> *[AVP]

Rule-Activation-Time AVP

Indicates when a rule is to be applied. It is sent by the PCRF in the Charging-Rule-Install AVP for each rule to which it applies.

Rule-Deactivation-Time AVP

Indicates when a rule is to be deactivated. It is sent by the PCRF in the Charging-Rule-Install AVP for each rule to which it applies.

Rule-Failure-Code AVP

This enumerated AVP is sent by the enforcement point to the enforcement point within the Charging-Rule-Report AVP. It identifies the reason a rule is being reported.

> 1 = UNKNOWN_RULE_NAME

Indicates a preprovisioned rule was not successfully activated because the rule identified in the Charging-Rule-Name or the Charging-Rule-Base-Name is not known by the enforcement point.

> 2 = RATING_GROUP_ERROR

Indicates the rule could not be activated or installed because of an invalid rating group identified in the Charging-Rule-Definition AVP.

> 3 = SERVICE_IDENTIFIER_ERROR

Indicates the rule could not be activated or installed because of an invalid service identifier identified in the Charging-Rule-Definition AVP.

4 = GW/enforcement point_MALFUNCTION

Indicates the rule could not be activated or installed because of a failure in the enforcement point.

5 = RESOURCES_LIMITATION

Indicates the rule could not be installed or activated because of resource limitations in the enforcement point.

6 = MAX_NR_BEARERS_REACHED

Indicates the rule could not be installed or activated because the maximum number of bearer channels has been allocated for the IP-CAN session.

7 = UNKNOWN_BEARER_ID

Indicates the rule could not be installed or activated because the Bearer-ID is invalid. This is only used in GPRS Networks where PCRF is performing bearer binding.

8 = MISSING_BEARER_ID

Indicates the rule could not be installed or activated because the Bearer-ID was not provided in the Charging-Rule-Install AVP.

9 = MISSING_FLOW_INFORMATION

Indicates the rule could not be installed or activated because the Flow-Information AVP was not provided in the Charging-Rule-Definition AVP.

10 = RESOURCE_ALLOCATION_FAILURE

Indicates the rule could not be installed or activated because the bearer was either released or failed.

11 = UNSUCCESSFUL_QOS_VALIDATION

Indicates the rule could not be installed or activated because QoS validation failed or the guaranteed bandwidth is greater than Max-Requested-Bandwidth.

12 = INCORRECT_FLOW_INFORMATION

Indicates the rule could not be installed or activated because the flow information provided by the PCRF is not supported.

13 = PS_TO_CS_HANDOVER

Indicates the rule could not be installed or activated because of a packet services to circuit-switched handover.

14 = TDF_APPLICATION_IDENTIFIER_ERROR

Indicates the rule could not be installed or activated because the Application-ID was invalid.

> 15 = NO_BEARER_BOUND

Indicates the rule could not be installed or activated because there is no bearer channel for the rule to be applied.

> 16 = FILTER_RESTRICTIONS

Indicates the rule could not be installed or activated because the Flow-Description cannot be handled by the enforcement point.

> 17 = AN_GW_FAILED

Indicates a failure at the AN gateway, and the PCRF should not send any rules to the enforcement point until the gateway is restored.

> 18 = MISSING_REDIRECT_SERVER_ADDRESS

Indicates the rule could not be installed or activated because the redirect server address provided in the Redirect-Server-Address AVP is invalid.

> 19 = CM_END_USER_SERVICE_DENIED

Indicates the charging system denied service because of service restrictions or limitations related to the subscriber.

> 20 = CM_CREDIT_CONTROL_NOT_APPLICABLE

Indicates service can be provided and no further credit control will be needed for the service.

> 21 = CM_AUTHORIZATION_REJECTED

Indicates the charging system denied service and is terminating the service for which credit is being requested.

> 22 = CM_USER_UNKNOWN

Indicates the subscriber indicated could not be found in the charging system.

> 23 = CM_RATING_FAILED

Indicates that the charging system cannot rate the service because of invalid AVP combinations or insufficient rating input.

Security-Parameter-Index AVP
This AVP contains the security parameter index for the IPsec packet.

Session-Release-Cause AVP
This is an enumerated AVP identifying the reason an IP-CAN was released by the PCRF.

> 0 = UNSPECIFIED_REASON

1 = UE_SUBSCRIPTION_REASON

2 = INSUFFICIENT_SERVER_RESOURCES

3 = IP_CAN_SESSION_TERMINATION

4 = UE_IP_ADDRESS_RELEASE

TDF-Application-Identifier AVP

This AVP is used to identify the application detection filter at the TDF. The application detection filter identifies an application and applies the associated rules. Applications can be identified by a list of URLs, for example.

TDF-Application-Instance-Identifier AVP

This identifier is created by the enforcement point for the correlation of application start and stop events to specific data flow descriptions. It is used when reporting from the enforcement point to the PCRF.

TDF-Destination-Host AVP

This AVP provides the destination host for the TDF.

TDF-Destination-Realm AVP

This AVP simply provides the destination realm for the TDF.

TDF-Information AVP

This grouped AVP is used to communicate details about the TDF that will be establishing a session with the PCRF and where application detection filters will be installed. The AVP is sent in the CCR command with the CC-Request-Type of INITIAL-REQUEST.

<TDF-Information>::=<AVP Header: 1087>

[TDF-Destination-Realm]

[TDF-Destination-Host]

[TDF-IP-Address]

TDF-IP-Address AVP

This AVP provides the IP address for the TDF.

TFT-Filter AVP

This AVP contains the data flow filter for one traffic flow template (TFT) filter. The following information is sent using this AVP:

- The protocol identifier for the next expected protocol header (default is IP)
- Direction (outbound)
- The action to be taken (set to permit)
- The source IP address
- The source and destination port (this can be a single port number, a list of ports, or ranges)
- Destination IP address

The Flow-Direction AVP is sent along with this AVP to indicate the direction the filter is to be applied.

TFT-Packet-Filter-Information AVP

This is a grouped AVP containing information from a single TFT packet filter. One AVP is sent for every rule request for a PDP context. The format is as follows:

<TFT-Packet-Filter-Information>::<AVP Header: 1013>

[Precedence]

[TFT-Filter]

[TOS-Traffic-Case]

[Security-Parameter-Index]

[Flow-Label]

*[AVP]

ToS-Traffic-Class AVP

The first octet of this AVP contains the IPv4 type of service or the IPv6 traffic class field and the second octet contains the ToS/traffic class mask field.

Tunnel-Header-Filter AVP

This AVP defines the outer header filter information for a mobile IP (MIP) tunnel. It includes

- The action "permit"
- Direction (in or out)
- Protocol
- Source IP address
- Source port for UDP tunneling
- Destination IP address
- Destination port for UDP tunneling

Tunnel-Header-Length AVP

This AVP indicates the length of the MIP tunnel header.

Tunnel-Information AVP

This is a grouped AVP sent from the enforcement point to the PCRF containing the header information from a single IP flow. It can also be sent by the PCRF to the BBERF.

<Tunnel-Information>::=<AVP Header: 1038>

[Tunnel-Header-Length]

2[Tunnel-Header-Filter]

*[AVP]

UDP-Source-Port AVP
This AVP contains the source port number when supporting interworking with a fixed broadband network.

UE-Local-IP-Address AVP
This AVP is used to provide the local IP address for the mobile device.

Usage-Monitoring-Information AVP
This is a grouped AVP that contains information used for usage monitoring. The volume and/or time of usage thresholds are identified in the Granted-Service-Unit AVP. The time that the threshold value is to be reapplied is contained in the Monitoring-Time AVP.

When reporting usage, the enforcement point sends the Used-Service-Unit AVP.

<Usage-Monitoring-Information>::=< AVP Header: 1067>

[Monitoring-Key]

0*2[Granted-Service-Unit]

0*2[Used-Service-Unit]

[Quota-Consumption-Time]

[Usage-Monitoring-Level]

[Usage-Monitoring-Report]

[Usage-Monitoring-Support]

*[AVP]

Usage-Monitoring-Level AVP
This enumerated AVP indicates how usage monitoring is to be applied; to the IP-CAN session or to a rule. If this AVP is not provided, the default value is PCC_RULE_LEVEL.

0 = SESSION_LEVEL

1 = PCC_RULE_LEVEL

2 = ADC_RULE_LEVEL

Usage-Monitoring-Report AVP
This enumerated AVP is sent by the PCRF to the enforcement point. It indicates whether or not accumulated usage is to be reported by the enforcement point, even if thresholds have not been reached.

0 = USAGE_MONITORING_REPORT_REQUIRED

Usage-Monitoring-Support AVP
This enumerated AVP is sent by the PCRF and indicates when usage monitoring is to be disabled.

0 = USAGE_MONITORING_DISABLED

AVP Name	AVP Code	Value Type
ADC-Rule-Base-Name	1095	UTF8String
ADC-Rule-Definition	1094	Grouped
ADC-Rule-Install	1092	Grouped
ADC-Rule-Name	1096	OctetString
ADC-Rule-Remove	1093	Grouped
ADC-Rule-Report	1097	Grouped

TABLE 7.7 AVPs Specific for the Sd Interface

User-Location-Info-Time AVP

This AVP reports the last known time where the mobile device was connected when the bearer channel was released. The information is sent to the PCRF, which then sends it to the P-CSCF.

Sd AVPs (Table 7.7)

ADC-Rule-Base-Name AVP

Similar to the ADC-Rule-Name, this AVP is used to identify a group of rules. The difference is that the ADC-Rule-Name identifies a single rule name, while this AVP is used to identify a group of rules.

ADC-Rule-Definition AVP

This AVP is used to define a rule being provisioned at the TDF. The specific rules are identified by the ADC-Rule-Name included in the description. Since this is used at the TDF, the traffic subject to the rule is application-based traffic, and identified using the TDF-Application-Identifier.

<ADC-Rule-Definition>::=<AVP Header: 1094>

{ADC-Rule-Name}

[TDF-Application-Identifier]

[Service-Identifier]

[Rating-Group]

[Reporting-Level]

[Online]

[Offline]

[Metering-Method]

[Precedence]

[Flow-Status]

[QoS-Information]

[Monitoring-Key]

[Redirect-Information]

[Mute-Notification]

[ToS-Traffic-Class]

*[AVP]

ADC-Rule-Install AVP

This grouped AVP is used for installing or activate rules at the TDF. The ADC-Rule-Definition AVP describes the parameters for the rule. If activating a rule then the ADC-Rule-Name AVP is used to identify the rule to be activated.

<ADC-Rule-Install>::=< AVP Header: 1092 >

*[ADC-Rule-Definition]

*[ADC-Rule-Name]

*[ADC-Rule-Base-Name]

[Monitoring-Flags]

[Rule-Activation-Time]

[Rule-Deactivation-Time]

*[AVP]

ADC-Rule-Name AVP

This AVP identifies the name of a specific rule. This is used for both PCRF defined rules and rules predefined at the TDF.

ADC-Rule-Remove AVP

This is a grouped AVP that is used to remove or deactivate rules at the TDF. The rules are identified using the ADC-Rule-Name AVP.

<ADC-Rule-Remove>::= <AVP Header: 1093>

*[ADC-Rule-Name]

*[ADC-Rule-Base-Name]

*[AVP]

ADC-Rule-Report AVP

This is a grouped AVP used to report the status of rules that cannot be installed or activated, defined by the PCRF and preconfigured at the TDF. The Rule-Failure-Code indicates why the rule installation or activation was not successful. There can be multiple instances of this AVP, especially where there are multiple rules within a group.

<ADC-Rule-Report>::=<AVP Header: 1097>

*[ADC-Rule-Name]

*[ADC-Rule-Base-Name]

[PCC-Rule-Status]

[Rule-Failure-Code]

[Final-Unit-Indication]

*[AVP]

CHAPTER 8

Connecting to IMS

When connecting to the IP Multimedia Subsystem (IMS) for voice services, the call session control functions (CSCF) in the IMS must access the Home Subscriber Server (HSS) prior to delivering any services. These services will be voice and multimedia, delivered through the packet network under the control of the IMS. It is important to note that the IMS provides control, while the bearer traffic is connected through the packet network.

The Cx interface is used to connect the HSS to the IMS (specifically the I-CSCF and the S-CSCF). This is to support the exchange of authentication and authorization information; authorization to use IMS services, authenticate users, and manage user data stored in the HSS and the IMS servers.

While both the Cx and the Dx are defined for connecting to the IMS, we will only be talking about the Cx specific procedures in this chapter. The Dx is used when the CSCF must connect to the Subscriber Location Function (SLF) to determine the correct HSS associated with a specific public user identity. The SLF provides the addresses of the HSS, and the CSCF must then send the message to the correct HSS.

When a Diameter proxy is used, this is not necessary. The CSCF sends its request to the proxy, which in turn locates the correct HSS and forwards the message. The use of a Diameter HSS proxy eliminates a lot of Diameter traffic in networks where multiple HSS are deployed.

The S-CSCF in the IMS is responsible for registration of a subscriber identity (or multiple identities). As the registrar, the S-CSCF must support specific capabilities, determined by the operator. In some networks, the S-CSCF is segregated throughout the network to support specific capabilities such as VPN support. The configuration is operator specific, but the I-CSCF must know how to steer subscribers to these S-CSCF implementations.

The I-CSCF bears the responsibility of assigning the S-CSCF to a public user identity, based on a number of factors, one being its capabilities. When a subscriber connects with the IMS for a specific service, and is routed from the P-CSCF to the I-CSCF, the I-CSCF must determine what capabilities the subscriber needs so it can determine which S-CSCF to assign to the subscriber.

The Interrogating-CSCF (I-CSCF) does not store this information, so it must go to the HSS each time to discover which Serving-CSCF (S-CSCF) supports these capabilities. This is done through the Cx interface.

In the IMS, the subscriber may use several identities. For each subscriber, there is a private user identity known only to the service provider. This is the identity used for billing. Each private user identity can have several public user identities. The public identities are used for different devices and different services.

For example a subscriber may use one public identity for email, and another for their iPhone. They may have yet another public identity for their iPad. The public identity can take the form of either an URI or a TEL URI. We will be talking about user identities throughout this chapter.

Cx Interface

The Cx interface as discussed earlier is used to connect between the I-CSCF, S-CSCF, and the HSS. The CSCFs access the HSS when a subscriber wishes to connect through the IMS (typically when they are making a VoLTE call, for example). This also applies to emergency calls through the packet network (VoLTE).

There are a number of options for supporting roaming in LTE networks, and a number of ideas around the support of voice calls in LTE networks. This book is not going to go into those technologies and how they work, but will instead stay focused on the Diameter interfaces and how they work.

Cx Procedures

Sessions at the Cx and Dx are considered as terminated, and therefore a termination request is not needed for these types of sessions. The Auth-Session-State AVP is set to the value of NO_STATE_MAINTAINED and sent in all commands.

Routing of commands on the Cx use the Destination-Host and Destination-Realm AVPs. The I-CSCF and the S-CSCF in the home IMS will know the host address for the HSS, but when sending commands to the HSS from outside the network it is recommended that a proxy be used. Routing should be based on the realm, and never the host address. This is to prevent abuse of interconnect.

When there are multiple HSS in a network, the I-CSCF and S-CSCF need to find the HSS associated with the subscriber. This is done through the Subscriber Location Function (SLF). The SLF will provide the addresses for the HSS in the network associated with a subscriber. This information is then returned to the CSCF, which in turn must then generate a request to the HSS.

The HSS address is included in the response. There can be multiple HSS addresses, provided in an ordered list. The CSCF then resends the request to the first HSS in the list, and if the CSCF does not receive a successful response, the CSCF sends the request to the next HSS in the list.

The I-CSCF or S-CSCF then sends the request to the first address in the list. If the first recipient of the request rejects the command, then the CSCF sends to the next HSS in the list. This continues until the CSCF reaches the right HSS. Note that usually the proxy will provide the correct HSS and repeating the request is not necessary.

Another option is to use a Diameter proxy. The request is sent to the proxy, which then determines which HSS is associated with the subscriber. The proxy then forwards the message to the proper HSS, thus eliminating additional signaling messages in the network.

When the S-CSCF receives a response from the HSS, the response will include the host and realm of the HSS. The S-CSCF stores the HSS host and realm address for each registered subscriber once it receives this information. Likewise, the HSS will store the host and realm address of the S-CSCF (derived from requests received by the S-CSCF). The I-CSCF is stateless and therefore does not store this information.

Command-Name	Abbreviation	Code
User-Authorization-Request	UAR	300
User-Authorization-Answer	UAA	300
Server-Assignment-Request	SAR	301
Server-Assignment-Answer	SAA	301
Location-Info-Request	LIR	302
Location-Info-Answer	LIA	302
Multimedia-Auth-Request	MAR	303
Multimedia-Auth-Answer	MAA	303
Registration-Termination-Request	RTR	304
Registration-Termination-Answer	RTA	304
Push-Profile-Request	PPR	305
Push-Profile-Answer	PPA	305

TABLE 8.1   Cx Commands and Their Codes

When the HSS, I-CSCF, and S-CSCF negotiate a session and perform the capabilities exchange, there are two possible values for the vendor identifier. Both 3GPP and ETSI are valid vendors in this case, and so the 3GPP value of 10415 and the ETSI value of 13019 will both be found, depending on the command and AVP.

Following are the commands used for Cx and a description of how they are used.

Cx Commands

There are six commands defined specifically for the Cx interface. You will not see these commands used on other interfaces. The header will carry the Application-ID to show the message is being sent via Cx. The Application-ID for Cx is 16777216. Table 8.1 identifies the commands used on Cx and their numeric codes.

User-Authorization-Request/Answer

The Diameter client (I-CSCF, for example) sends the User-Authorization-Request (UAR) command to the Diameter server (the HSS in this case) when a device is registering for an IMS session. The I-CSCF will query the HSS to verify permissions for a received public user identity.

The I-CSCF also checks the HSS to verify the public user identity is associated with the private user identity received as a security check (authentication). The I-CSCF will also receive the address of the S-CSCF where this public user identity is stored.

When the HSS receives the UAR, the HSS will first check to make sure the user identity is known to the HSS. If the HSS does not know the user identity, it will return the User-Authorization-Answer (UAA) with an Experimental-Result-Code value of DIAMETER_ERROR_USER_UNKNOWN. If both the public and the private user identity are stored in the HSS, the HSS will then verify the private user identity is associated with the public user identity. If not, the HSS sends the UAA with the Experimental-Result-Code value DIAMETER_ERROR _IDENTITIES_DONT_MATCH.

When the HSS has determined the user identity is valid and is stored in the receiving HSS, it then checks the permissions for the user identity. If the user identity is not allowed access to the IMS, then the HSS returns the UAA with the Result-Code value of DIAMETER_AUTHORIZATION_REJECTED. Authorization then fails and the mobile device is prevented from accessing the IMS, unless this is an emergency session. If the UAR-Flags indicates this is an emergency session, the HSS will not perform any further checks and the IMS connection is granted, but limited to the emergency APN.

The HSS next checks the User-Authorization-Type AVP to determine if this is a registration request. If it is a registration request, then the HSS will verify the user identity for permissions. If the mobile device is in a visited network roaming, the HSS must also determine if roaming is allowed. If the user identity is allowed roaming in the visited network, the HSS checks to see if the user identity is registered.

Registration is performed through the S-CSCF. If the user identity has registered in the IMS, then the HSS will return the name of the S-CSCF where the registration occurred. Since the user identity is already registered, the HSS will send the UAA with Experimental-Result-Code value of DIAMETER_SUBSEQUENT_REGISTRATION.

Note that a subscriber can have multiple public user identities. The subscriber can register multiple public user identities simultaneously using implicit registration. If the S-CSCF needs to make a change to an identity that has been implicitly registered as part of a group of identities, the change is applied to all of the public user identities that the subscriber registered.

The HSS will send service information for a group of identities that were registered by the subscriber using implicit registration. Each public user identity that is part of a group is listed in the HSS response. The S-CSCF stores this list, and selects the first identity in the group to be the default for the subscriber registration. The user profile for each of the public user identities is also sent in the response.

If the S-CSCF sends deregistration for a private or public user identity, and the identity belongs to a group, then all of the identities in the group are deregistered. This change is implemented in both the HSS and the S-CSCF.

If the public user identity is not registered, a S-CSCF will not be assigned and the HSS will not be able to find the S-CSCF name for the identity. In this case, the HSS will send the UAA to the I-CSCF with the Server-Capabilities AVP, allowing the I-CSCF the ability to assign the S-CSCF to the public user identity. The Server-Capabilities AVP contains the requirements for the IMS subscription. If this is not included in the UAA, it indicates that the I-CSCF can choose any S-CSCF. The Experimental-Result-Code value for the response will be DIAMETER_FIRST_REGISTRATION.

The User-Authorization-Request (UAR) command is structured as follows:

<User-Authorization-Request>::=<Diameter Header: 300, REQ, PXY, 16777216>

<Session-ID>

{Vendor-Specific-Application-ID}

{Auth-Session-State}

{Origin-Host}

{Origin-Realm}

[Destination-Host]

{Destination-Realm}

{User-Name}

[OC-Supported-Features]

*[Supported-Features]

{Public-Identity}

{Visited-Network-Identifier}

[User-Authorization-Type]

[UAR-Flags]

*[Proxy-Info]

*[Route-Record]

*[AVP]

The User-Authorization-Answer (UAA) command is used in response to the request and is structured as follows:

< User-Authorization-Answer>::=<Diameter Header: 300, PXY, 16777216>

<Session-ID>

{Vendor-Specific-Application-ID}

[Result-Code]

[Experimental-Result]

{Auth-Session-State}

{Origin-Host}

[OC-Supported-Features]

[OC-OLR}

*[Supported-Features]

[Server-Name]

[Server-Capabilities]

*[Failed-AVP]

*[Proxy-Info]

*[Route-Record]

*[AVP]

Server-Assignment-Request/Answer

The Server-Assignment-Request (SAR) command is used by the S-CSCF to communicate to the HSS when it has been assigned to a public user identity, or when it needs to clear a public user identity from the HSS. The S-CSCF also uses this command to obtain subscriber profile information, such as permissions and charging information.

The HSS maintains information about the state of all subscriber IMS registrations. This includes the S-CSCF name. The S-CSCF providing services to the registered subscriber sends updates to the HSS regarding the registration status of the subscriber's user identity (both private and public).

When the HSS receives a request from the S-CSCF, the HSS checks to verify the public user identity is stored in the HSS, and that the private user identity and the public user identity are associated with one another in the HSS. If the two are not associated, the HSS will send the SAA with the Experimental-Result-Code value of DIAMETER_ERROR_IDENTITIES_DONT_MATCH. Note that a subscription can have multiple public user identities, so the HSS must check for all known public user identities for a subscription. If the public user identity is shown as registered in the HSS to an S-CSCF other than the one sending the request, the HSS will send the response with the name of the S-CSCF "of record" and the Experimental-Result-Code value of DIAMETER_ERROR_IDENTITY_ALREADY_REGISTERED.

The HSS will check the registration state for all public user identities provided in the request, but in the event there are no public user identities provided, the HSS will check all the public user identities associated with the provided private user identities. If the request is to deregister the private user identity, and the HSS locates a public user identity registered, it will change the status to deregistered and clear the S-CSCF name from the profile.

In the event the public user identity is associated with more than one private user identity, the HSS will check the registration state of the public user identity provided. If the request is to deregister, the public user identity will be deregistered for the one private user identity.

The HSS can overwrite the S-CSCF listed as last registered provided IMS restoration procedures are supported and the reassignment pending flag is set in the request. The flag is reset in the response message.

When the S-CSCF sending the request and is stored in the HSS as the assigned S-CSCF, and the User-Data-Already-Available AVP is set to the value of USER_DATA_NOT_AVAILABLE, the HSS takes the user data provided in the request and stores it with the subscriber profile. If the User-Data-Already-Available AVP is set to the value USER_DATA_ALREADY_AVAILABLE, the HSS can still download the subscriber profile but it will not send any subscriber data in its response. If there are multiple private user identities for an IMS subscription, the HSS includes the other identities in the Associated-Identities AVP in its response.

If the HSS is unable to support the request from the S-CSCF, it shall send a Result-Code of DIAMETER_UNABLE_TO_COMPLY in the response. The format for this command is as follows:

< Server-Assignment-Request>::=<Diameter Header: 301, REQ, PXY, 16777216>

<Session-ID>

{Vendor-Specific-Application-ID}

{Auth-Session-State}

{Origin-Host}

{Origin-Realm}

[Destination-Host]

{Destination-Realm}

[User-Name]

[OC-Supported-Features]

*[Supported-Features]

*[Public-Identity]

[Wildcarded-Public-Identity]

{Server-Name}

{Server-Assignment-Type}

{User-Data-Already-Available}

[SCSCF-Restoration-Info]

[Multiple-Registration-Indication]

[Session-Priority]

*[Proxy-Info]

*[Route-Record]

*[AVP]

The Server-Assignment-Answer (SAA) command is sent in response and if there is no error response, the contents will provide the information that the S-CSCF needs to provide services to the user.

< Server-Assignment-Answer>::=<Diameter Header: 301, PXY, 16777216>

<Session-ID>

{Vendor-Specific-Application-ID}

[Result-Code]

[Experimental-Result]

{Auth-Session-State}

{Origin-Host}

{Origin-Realm}

[User-Name]

[OC-Supported-Features]

[OC-OLR]

*[Supported-Features]

[User-Data]

[Charging-Information]

[Associated-Identities]

[Loose-Route-Indication]

*[SCSCF-Restoration-Info]

[Associated-Registered-Identities]

[Server-Name]

[Wildcarded-Public-Identity]

[Privileged-Sender-Indication]

*[Failed-AVP]

*[Proxy-Info]

*[Route-Record]

*[AVP]

Location-Info-Request/Answer

The Location-Info-Request (LIR) command is sent by the I-CSCF to the HSS when looking for the name of the assigned S-CSCF for a specific public user identity. The command is sent for each public user identity sent by the I-CSCF.

If the public user identity sent in the request is stored in the HSS, and it is shown as registered, then the HSS will return the stored S-CSCF user name contained in the Server-Name AVP. The Server-Name AVP contains the SIP URI for the assigned S-CSCF.

If the provided public user identity is not registered, the HSS will check for any other public user identities associated with the same IMS subscription, and if there are any other public user identities registered, the HSS will provide the S-CSCF assigned to this identity.

The format for this command is as follows:

<Location-Info-Request>::=<Diameter Header: 302, REQ, PXY, 16777216>

<Session-ID>

{Vendor-Specific-Application-ID}

{Auth-Session-State}

{Origin-Host}

{Origin-Realm}

[Destination-Host]

{Destination-Realm}

[Originating-Request]

[OC-Supported-Features]

*[Supported-Features]

{Public-Identity}

[User-Authorization-Type]

[Session-Priority]

*[Proxy-Info]

*[Route-Record]

*[AVP]

The Diameter server in response to the request sends the LIA command.

<Location-Info-Answer>::=<Diameter Header: 302, PXY, 16777216>

<Session-ID>

{Vendor-Specific-Application-ID}

[Result-Code]

[Experimental-Result]

{Auth-Session-State}

{Origin-Host}

{Origin-Realm}

[OC-Supported-Features]

[OC-OLR]

*[Supported-Features]

[Server-Name]

[Server-Capabilities]

[Wildcarded-Public-Identity]

[LIA-Flags]

*[Failed-AVP]

*[Proxy-Info]

*[Route-Record]

*[AVP]

Multimedia-Auth-Request Command

When a subscriber is connecting to the IMS, the S-CSCF must query the HSS using the Multimedia-Auth-Request (MAR) to retrieve the authentication information. The HSS provides the authentication vectors to the S-CSCF using the MAA command.

When the HSS receives the MAR, it first checks the public and private identities sent in the MAR to ensure they are associated with the HSS. Then the HSS will check the authentication scheme to make sure it is able to support the type of authentication requested.

If these check out okay, the HSS then checks the registration status of the public identities provided. If the public identity is registered, then the S-CSCF name stored in the HSS should match the name of the S-CSCF provided in the request. If the request provides a different name, the HSS will change the S-CSCF name to the new name received in the request.

The format for the MAR command is as follows:

<Multimedia-Auth-Request>::=<Diameter Header: 303, REQ, PXY, 16777216>

<Session-ID>

{Vendor-Specific-Application-ID}

{Auth-Session-State}

{Origin-Host}

{Origin-Realm}

{Destination-Realm}

[Destination-Host]

{User-Name}

[OC-Supported-Features]

*[Supported-Features]

{Public-Identity}

{SIP-Auth-Data-Item}

{SIP-Number-Auth-Items}

{Server-Name}

*[Proxy-Info]

*[Route-Record]

*[AVP]

The server as a response sends the authentication information requested by the client using the MAA command. The format for the response is as follows:

< Multimedia-Auth-Answer>::=<Diameter Header: 303, PXY, 16777216>

<Session-ID>

{Vendor-Specific-Application-ID}

[Result-Code]

[Experimental-Result]

{Auth-Session-State}

{Origin-Host}

{Origin-Realm}

[User-Name]

[OC-Supported-Features]

[OC-OLR]

*[Supported-Features]

[Public-Identity]

[SIP-Number-Auth-Items]

*[SIP-Auth-Data-Item]

*[Failed-AVP]

*[Proxy-Info]

*[Route-Record]

*[AVP]

Registration-Termination-Request/Answer

The Registration-Termination-Request (RTR) command is sent by the HSS to the S-CSCF when it needs to deregister a subscriber. The S-CSCF then removes all associated information for the subscriber from its storage.

The HSS can deregister one identity or many identities, or it can deregister one private user identity and all of its public user identities. When this occurs the HSS will also send the Reason-Code of PERMANENT_TERMINATION, SERVER_CHANGE, or REMOVE_S-CSCF.

The Deregistration-Reason AVP is used to identify why the HSS is sending deregistration to the S-CSCF. Public user identities that are registered for emergency IMS sessions are not deregistered.

The format for this command is as follows:

< Registration-Termination-Request>::=<Diameter Header: 304, REQ, PXY, 16777216>

<Session-ID>

{Vendor-Specific-Application-ID}

{Auth-Session-State}

{Origin-Host}

{Origin-Realm}

{Destination-Host}

{Destination-Realm}

{User-Name}

[Associated-Identities]

*[Supported-Features]

*[Public-Identity]

{Deregistration-Reason}

*[Proxy-Info]

*[Route-Record]

*[AVP]

The Registration-Termination-Answer (RTA) command is sent in response to the request and is formatted as follows:

< Registration-Termination-Answer>::=<Diameter Header: 304, PXY, 16777216>

<Session-ID>

{Vendor-Specific-Application-ID}

[Result-Code]

[Experimental-Result]

{Auth-Session-State}

{Origin-Host}

{Origin-Realm}

[Associated-Identities]

*[Supported-Features]

*[Identity-With-Emergency-Registration]

*[Failed-AVP]

*[Proxy-Info]

*[Route-Record]

*[AVP]

Push-Profile-Request/Answer

The Push-Profile-Request (PPR) command is sent by HSS to update subscription data and SIP digest authentication for an IMS subscriber in the S-CSCF. The S-CSCF uses this command to send the user profile, charging information, and SIP Digest authentication data to the HSS.

If a new public user identity associated with an existing user is being added in the HSS, the HSS updates the user profile and sends to the S-CSCF. The newly added public user identity is added to an existing private user identity for the same subscriber, and will assume the same state as the existing private user identity (registered or unregistered). This is also used when removing public user identities from a subscription profile.

When changing the charging information associated with a subscriber profile, the private user identity must be provided in the PPR command. The HSS will push this information to the S-CSCF if the public user identity(s) for the subscription is registered or unregistered.

The same is true when changing the SIP Digest authentication data. The private user identity is required to update the change in the S-CSCF. Any registered public user identity whose authentication information changes will be sent to the S-CSCF immediately.

There can be multiple private user identities associated with a public user identity. In this case, the HSS has to select one of the private user identities and include it in the command. Only one request is sent for the public user identity.

The format of this command is as follows:

<Push-Profile-Request>::=<Diameter Header: 305, REQ, PXY, 16777216>

< Session-ID>

{Vendor-Specific-Application-ID}

{Auth-Session-State}

{Origin-Host}

{Origin-Realm}

{Destination-Host}

{Destination-Realm}

{User-Name}

*[Supported-Features]

[User-Data]

[Charging-Information]

[SIP-Auth-Data-Item]

*[Proxy-Info]

*[Route-Record]

*[AVP]

The Diameter client sends the PPA command in response, using this format:

<Push-Profile-Answer>::= < Diameter Header: 305, PXY, 16777216>

<Session-ID>

{Vendor-Specific-Application-ID}

[Result-Code]

[Experimental-Result]

{Auth-Session-State}

{Origin-Host}

{Origin-Realm}

*[Supported-Features]

*[Failed-AVP]

*[Proxy-Info]

*[Route-Record]

*[AVP]

Cx AVPs

The following section provides descriptions of the AVPs defined specifically for the Cx interface. These AVPs are specific to the Cx and are not reused on other interfaces. For the purpose of brevity, the reused AVPs are not included.

There are some AVPs that are mapped from the Cx messaging (non-Diameter) when Diameter is implemented on Cx. Table 8.2 shows the mapping between Cx messaging and Diameter.

The Diameter AVPs defined specifically for the Cx interface are shown in numerical order in Table 8.3.

The Diameter Cx AVPs are described in alphabetical order below to make it easier to find the definition. The description for the commands above will provide more information as to how some of these AVPs may be used. For a complete definition of how these AVPs are used in all situations, refer to the 3GPP standards.

Associated-Registered-Identities AVP   This grouped AVP contains the private user identities that are registered with the public user identity provided in the request.

<Associated-Registered-Identities>::=<AVP Header: 647 10415>

*[User-Name]

*[AVP]

Cx Parameter	AVP Name
Visited Network Identifier	Visited-Network-Identifier
Public Identity	Public-Identity
Private Identity	User-Name
S-CSCF Name	Server-Name
AS Name	
S-CSCF capabilities	Server-Capabilities
Result	Result-Code
	Experimental-Result-Code
User profile	User-Data
Server Assignment Type	Server-Assignment-Type
Authentication data	SIP-Auth-Data-Item
Item Number	SIP-Item-Number
Authentication Scheme	SIP-Authentication-Scheme
Authentication Information	SIP-Authenticate
Authorization Information	SIP-Authorization
Confidentiality Key	Confidentiality-Key
Integrity Key	Integrity-Key
Number Authentication Items	SIP-Number-Auth-Items
Reason for deregistration	Deregistration-Reason
Charging Information	Charging-Information
Routing Information	Destination-Host
Type of Authorization	Authorization-Type
Associated Private Identities	Associated-Identities
Digest Authenticate	SIP-Digest-Authenticate
Digest Realm	Digest-Realm
Digest Algorithm	Digest-Algorithm
Digest QoP	Digest-QoP
Digest HA1	Digest-HA1
Line Identifier	Line-Identifier
Wildcarded Public Identity	Wildcarded-Public Identity
Loose-Route Indication	Loose-Route-Indication
S-CSCF Restoration Information	SCSCF-Restoration-Info
Multiple Registration Indication	Multiple-Registration-Indication
Priviledged-Sender Indication	Priviledged-Sender-Indication
LIA Flags	LIA-Flags

TABLE 8.2   Mapping between Cx Messages and Diameter on Cx

Attribute Name	AVP Code	Value Type
Visited-Network-Identifier	600	OctetString
Public-Identity	601	UTF8String
Server-Name	602	UTF8String
Server-Capabilities	603	Grouped
Mandatory-Capability	604	Unsigned32
Optional-Capability	605	Unsigned32
User-Data	606	OctetString
SIP-Number-Auth-Items	607	Unsigned32
SIP-Authentication-Scheme	608	UTF8String
SIP-Authenticate	609	OctetString
SIP-Authorization	610	OctetString
SIP-Authentication-Context	611	OctetString
SIP-Auth-Data-Item	612	Grouped
SIP-Item-Number	613	Unsigned32
Server-Assignment-Type	614	Enumerated
Deregistration-Reason	615	Grouped
Reason-Code	616	Enumerated
Reason-Info	617	UTF8String
Charging-Information	618	Grouped
Primary-Event-Charging-Function-Name	619	DiameterURI
Secondary-Event-Charging-Function-Name	620	DiameterURI
Primary-Charging-Collection-Function-Name	621	DiameterURI
Secondary-Charging-Collection-Function-Name	622	DiameterURI
User-Authorization-Type	623	Enumerated
User-Data-Already-Available	624	Enumerated
Confidentiality-Key	625	OctetString
Integrity-Key	626	OctetString
Supported-Features	628	Grouped
Feature-List-ID	629	Unsigned32
Feature-List	630	Unsigned32
Supported-Applications	631	Grouped
Associated-Identities	632	Grouped
Originating-Request	633	Enumerated
Wildcarded-Public-Identity	634	UTF8String
SIP-Digest-Authenticate	635	Grouped
UAR-Flags	637	Unsigned32
Loose-Route-Indication	638	Enumerated

TABLE 8.3   AVPs Used on the Cx Interface

Attribute Name	AVP Code	Value Type
SCSCF-Restoration-Info	639	Grouped
Path	640	OctetString
Contact	641	OctetString
Subscription-Info	642	Grouped
Call-ID-SIP-Header	643	OctetString
From-SIP-Header	644	OctetString
To-SIP-Header	645	OctetString
Record-Route	646	OctetString
Associated-Registered-Identities	647	Grouped
Multiple-Registration-Indication	648	Enumerated
Restoration-Info	649	Grouped
Session-Priority	650	Enumerated
Identity-with-Emergency-Registration	651	Grouped
Priviledged-Sender-Indication	652	Enumerated
LIA-Flags	653	Unsigned32
OC-Supported-Features	TBD	Grouped
OC-OLR	TBD	Grouped
Initial-CSeq-Sequence-Number	654	Unsigned32
SAR-Flags	655	Unsigned32

TABLE 8.3   AVPs Used on the Cx Interface (*Continued*)

Associated-Identities AVP   This grouped AVP provides the private user identities associated with an IMS subscription.

    <Associated-Identities>::=<AVP Header: 632 10415>

    *[User-Name]

    *[AVP]

Call-ID-SIP-Header AVP   This AVP provides the content of the Call-ID SIP header and is used in the Subscription-Info AVP.

Charging-Information AVP   This grouped AVP contains the address of charging functions.

    <Charging-Information>::=<AVP Header: 618 10415>

    [Primary-Event-Charging-Function-Name]

    [Secondary-Event-Charging-Function-Name]

    [Primary-Charging-Collection-Function-Name]

    [Secondary-Charging-Collection-Function-Name]

    *[AVP]

Confidentiality-Key AVP This AVP is used by the HSS for sending the confidentiality key (CK) used in authentication.

Contact AVP This AVP contains the SIP URI as found in the SIP header of the same name. The Contact address is used in SIP to provide the address for a subscriber device to be used in subsequent requests and responses.

Deregistration-Reason AVP This is a grouped AVP used to indicate why deregistration was requested.

<Deregistration-Reason>::=<AVP Header: 615 10415>

{Reason-Code}

[Reason-Info]

*[AVP]

Digest-Algorithm AVP This AVP is used as part of the Digest authentication and its use can be found in RFC 4740.

Digest-HA1 AVP This AVP is used as part of the Digest authentication and its use can be found in RFC 4740.

Digest-QoP AVP This AVP is used as part of the Digest authentication and its use can be found in RFC 4740.

Digest-Realm AVP This AVP is used to provide the Digest realm used in authentication. The full description of how this AVP is used can be found in RFC 4740.

Feature-List AVP The Feature-List AVP is used to identify the features supported by the destination host. This is a bit-masked AVP, with the bits defined as follows (Table 8.4):

Feature-List-ID AVP This is used in the Supported-Features AVP for identifying a list of features being sent.

Framed-Interface-ID AVP This AVP contains the IPv6 interface identifier that is to be configured for the user.

Framed-IP-Address AVP This AVP contains the IPv4 address but can also be used to request an address. There are two special values according to RFC 4005:

0xFFFFFFFF indicates that the NAS is to allow the user to select an IP address

0xFFFFFFFE indicates that the NAS is to select an address for the user

Bit	Feature
0	Shared initial Feature Criteria (SiFC)
1	Alias Indication
2	IMS Restoration Indication
3	P-CSCF Restoration Mechanism

TABLE 8.4 Cx Feature List Definitions

Framed-IPv6-Prefix AVP   This is similar to the Framed-IP-Address but applies to IPv6 addresses. The format for the content of this AVP is found in RFC 3162.

From-SIP-Header AVP   This AVP contains the content from the From SIP header and is used in the Subscription-Info AVP.

Identity-with-Emergency-Registration AVP   This is a grouped AVP containing a private and a public user identity that are both emergency registered.

< Identity-with-Emergency-Registration>::=<AVP Header: 651 10415>

{User-Name}

{Public-Identity}

*[AVP]

Integrity-Key AVP   This AVP is used by the HSS to send the integrity key (IK) used in authentication.

LIA-Flags AVP   This AVP is bit-masked and when the bit is set indicates the HSS is to return the AS name in the Server-Name AVP (Table 8.5).

Line-Identifier AVP   The Line-Identifier AVP identifies an access line that is associated with a fixed broadband subscriber. The Vendor ID is ETSI (13019).

Loose-Route-Indication AVP   This is an enumerated AVP used to indicate if the S-CSCF needs to use loose routing when serving the registered users.

0 = LOOSE_ROUTE_NOT_REQUIRED

1 = LOOSE_ROUTE_REQUIRED

Mandatory-Capability AVP   This AVP indicates the capabilities that are required of the S-CSCF in the IMS. Each of the values can be individual requirements or a set of requirements. The HSS sends this to the I-CSCF and can be used to distribute subscribers between different S-CSCFs based on specific capabilities. A capability may be geography, specific feature sets, or maybe a specific role being provided by the S-CSCF (service provider defined). For example, a S-CSCF may be designated to handle priority and preemption for first responders. It is the services providers' responsibility to define a value for each of the capabilities (Table 8.6).

Multiple-Registration-Indication AVP   This enumerated AVP indicates whether or not this request is part of a multiple registration.

0 = NOT_MULTIPLE_REGISTRATION

1 = MULTIPLE_REGISTRATION

Bit	Description
0	PSI Direct Routing Indication

TABLE 8.5   LIA-Flags Value for Cx

Mandatory Capabilities	Description
Wildcarded PSI	The S-CSCF must support wildcard PSIs
OrigUnreg SPT	The S-CSCF must be able to support initial filter criteria with a value of "Originating_Unregistered" received by the HSS
OrigCDIV SPT	The S-CSCF must support initial filter criteria with the value "Originating_CDIV" received by the HSS
SIP Digest Authentication	The S-CSCF must support SIP digest authentication
NASS Bundled Authentication	The S-CSCF must support NASS bundling authentication
Wildcarded IMPUs	The S-CSCF must support wildcard IP multimedia public identities (IMPUs)
Loose-Route	The S-CSCF must support loose-route
Service Level Trace	The S-CSCF must support service level trace
Priority Service	The S-CSCF must support a network pre-configured default namespace and the service priority level associated with that namespace
Extended Priority	The S-CSCF must support priority namespaces and their associated priority levels.
Early IMS Security	The S-CSCF must support GPRS IMS bundled authentication (GIBA)
Reference Location	The S-CSCF must support reference location
Privileged-Sender	The S-CSCF must support privileged sender
IMSI	The S-CSCF must support the processing of received messages with IMSI
Maximum Number of allowed simultaneous registrations	The S-CSCF must support processing of maximum number of allowed simultaneous registrations per user

TABLE 8.6   Mandatory Capabilities Suggested by 3GPP

Optional-Capability AVP   The HSS can also identify optional capabilities that a S-CSCF must support (see also Mandatory-Capabilities). It is the services providers' responsibility to define a value for each of the capabilities (Table 8.7).

Originating-Request AVP   This enumerated AVP simply identifies the originator of a SIP request, in this case an application server in the SIP domain.

0 = Originating

Capability	Description
Shared iFC sets	The S-CSCF may support shared initial filter criteria
Display Name	The S-CSCF may support display name
Alias	The S-CSCF may support the AliasInd feature

TABLE 8.7   Optional Capabilities Suggested by 3GPP

Path AVP   This AVP contains a comma-delimited list of SIP proxies using the syntax defined in RFC 3327.

```
Path = "Path" HCOLON path-value *( COMMA path-value )
path-value = name-addr *( SEMI rr-param )
```

Primary-Charging-Collection-Function-Name AVP   The primary charging collection function is where billing records are to be sent for time-based or volume-based charges. The address is in URI format, and is used to also create the Destination-Host and Destination-Realm AVPs in subsequent charging messages.

Primary-Event-Charging-Function-Name AVP   In prepaid charging networks, the primary online charging function must be identified so events can be recorded in billing. This AVP provides the address for the primary charging function for routing of these event messages. The address is also used for the Destination-Host AVP in subsequent messages carrying charging information. The format is the URI for the platform.

Privileged-Sender-Indication AVP   This is an enumerated AVP used to indicate that the identity provided is a privileged sender. Note that the spelling for this AVP (Priviledged) is incorrect in the standard, and therefore could be incorrectly spelled in the protocol implementation as well.

0 = NOT_PRIVILEDGED_SENDER

1 = PRIVILEDGED_SENDER

Public-Identity AVP   The IMS public identity of a user is provided in this AVP using the SIP URI or the SIP TEL format.

Reason-Code AVP   This enumerated AVP is used to define why network initiated deregistration was requested.

0 = Permanent termination

1 = New server assigned

2 = Server change

3 = Remove the S-CSCF

Reason-Info AVP   This AVP contains information about the reason for deregistration. The information is used to inform the subscriber. The following values can be used:

PERMANENT_TERMINATION

NEW_SERVER_ASSIGNED

SERVER_CHANGE

REMOVE_S-CSCF

Record-Route AVP   This AVP contains the content of the Record-Route SIP header and is used in the Subscription-Info AVP. The Record-Route headers are provided as a list separated by commas in this AVP.

Restoration-Info AVP This grouped AVP contains the data for a specific registration to be used by the S-CSCF to deliver services to the subscriber.

<Restoration-Info>::=<AVP Header: 649 10415>

{Path}

{Contact}

[Initial-CSeq-Sequence-Number]

[Call-ID-SIP-Header]

[Subscription-Info]

*[AVP]

Result-Code AVP There are a number of result code values defined specifically for the Cx interface. Those values are provided here.

SUCCESSFUL VALUES

DIAMETER_FIRST_REGISTRATION (2001)

DIAMETER_SUBSEQUENT_REGISTRATION (2002)

DIAMETER_UNREGISTERED_SERVICE (2003)

DIAMETER_SUCCESS_SERVER_NAME_NOT_STORED (2004)

It should be noted that the S-CSCF is assigned to the subscriber when the value 2001 is sent.

PERMANENT FAILURES

DIAMETER_ERROR_USER_UNKNOWN (5001)

DIAMETER_ERROR_IDENTITIES_DONT_MATCH (5002)

DIAMETER_ERROR_IDENTITY_NOT_REGISTERED (5003)

DIAMETER_ERROR_ROAMING_NOT_ALLOWED (5004)

DIAMETER_ERROR_IDENTITY_ALREADY_REGISTERED (5005)

DIAMETER_ERROR_AUTH_SCHEME_NOT_SUPPORTED (5006)

DIAMETER_ERROR_IN_ASSIGNMENT_TYPE (5007)

DIAMETER_ERROR_TOO_MUCH_DATA (5008)

DIAMETER_ERROR_NOT_SUPPORTED_USER_DATA (5009)

DIAMETER_ERROR_FEATURE_UNSUPPORTED (5011)

DIAMETER_ERROR_SERVING_NODE_FEATURE_UNSUPPORTED (5012)

SAR-Flags AVP This bit-masked AVP indicates that the P-CSCF restoration mechanism feature is to be executed when this bit is set. It is only used when the Server-Assignment-Type AVP contains the value of ADMINISTRATIVE_DEREGISTATION or UNREGISTERED_ USER (Table 8.8).

Bit	Description
0	P-CSCF Restoration Indication

TABLE 8.8 SAR-Flags Value for Cx

Session-Priority AVP   This is an enumerated AVP indicating the session priority. It will usually be placed close to the header for optimal processing. The definition of each priority is operator dependent, but emergency sessions should always be given the highest priority (priority 0).

0 = PRIORITY-0

1 = PRIORITY-1

2 = PRIORITY-2

3 = PRIORITY-3

4 = PRIORITY-4

SCSCF-Restoration-Info AVP   This grouped AVP provides information needed by the S-CSCF for processing requests from users.

<SCSCF-Restoration-Info>::=<AVP Header: 639 10415>

{User-Name}

1*{Restoration-Info}

[SIP-Authentication-Scheme]

*[AVP]

Secondary-Charging-Collection-Function-Name AVP   The secondary charging function address is the destination for time and volume-based charging records to be sent. This AVP provides the URI, which is also used for creating the Destination-Host and Destination-Realm AVPs.

Secondary-Event-Charging-Function-Name AVP   Like the Primary-Event-Charging-Function-Name AVP, this one identifies the URI for the charging platform acting as the secondary charging function in the prepaid network. The address is also used to create the Destination-Host and Destination-Realm AVPs in subsequent messages carrying event information to the charging function.

Server-Assignment-Type AVP   This is an enumerated AVP identifying the type of server update, request or modification to be performed. It is part of a Server-Assignment-Request.

0 = No assignment—used to request a user profile from the HSS without affecting the registration state of the user.

1 = Registration—the subscriber has connected in the IMS with their first registration, generating this request

2 = Reregistration—the subscriber has reregistered with the IMS

3 = Unregistered user—a request was received by the S-CSCF for a subscriber that is not registered

4 = Timeout deregistration—a SIP registration timer has expired for a specific subscriber

5 = User deregistration—a subscriber initiated a deregistration at the S-CSCF

6 = Timeout deregistration store server name—the HSS is being required to store the S-CSCF name when a SIP registration timer expires for a subscriber

7 = User deregistration store server name—the HSS is being required to store the S-CSCF name when a subscriber deregisters from the S-CSCF

8 = Administrative deregistration—the S-CSCF has deregistered a subscriber because of a network failure or because of administrative reasons

9 = Authentication failure—subscriber authentication failed

10 = Authentication timeout—there was a timeout during authentication

11 = Deregistration too much data—the S-CSCF has received more information than it can accept when it requested a user profile

12 = AAA user data request—not used

13 = PGW update—not used

14 = Restoration—not used

Server-Capabilities AVP This is a grouped AVP with information used by the I-CSCF when selecting the S-CSCF for a subscriber.

<Server-Capabilities>::=<AVP Header: 603 10415>

*[Mandatory-Capability]

*[Optional-Capability]

*[Server-Name]

*[AVP]

Server-Name AVP This AVP provides the SIP-URL of the S-CSCF or other IMS SIP server. The Server-Name can be used to steer a group of subscribers to a specific S-CSCF. The HSS sends this AVP in the Server-Capabilities AVP, instead of the Mandatory-Capabilities AVP.

SIP-Auth-Data-Item AVP This is a grouped AVP containing authentication and authorization information.

<SIP-Auth-Data-Item>::=<AVP Header: 612 10415>

[SIP-Item-Number]

[SIP-Authentication-Scheme]

[SIP-Authenticate]

[SIP-Authorization]

[SIP-Authentication-Context]

[Confidentiality-Key]

[Integrity-Key]

[SIP-Digest-Authenticate]

[Framed-IP-Address]

[Framed-IPv6-Prefix]

[Framed-Interface-ID]

*[Line-Identifier]

*[AVP]

SIP-Authenticate AVP In this AVP, specific parts of the WWW-Authenticate or Proxy-Authenticate SIP headers needed for a SIP response are identified. The formatting of the content is defined in 3GPP TS 29.228.

SIP-Authentication-Context AVP This AVP contains information related to the authentication of a SIP user. The information is usually not part of the SIP authentication headers.

SIP-Authentication-Scheme AVP In this AVP, the authentication used in the SIP sessions is provided as defined in 3GPP TS 29.228.

SIP-Authorization AVP This AVP provides the specific parts of the WWW-Authenticate or Proxy-Authenticate SIP headers that can be included in a SIP request. See 3GPP TS 29.228 for exact formatting of this content.

SIP-Digest-Authenticate AVP This grouped AVP contains the content from the SIP WWW-Authenticate or Proxy-Authenticate headers.

<SIP-Digest-Authenticate>::=<AVP Header: 635 10415>

{Digest-Realm}

[Digest-Algorithm]

{Digest-QoP}

{Digest-HA1}

*[AVP]

SIP-Item-Number AVP This AVP is used where the SIP-Auth-Data-Item contains multiple SIP-Auth-Data-Item AVPs and provides the order in which these AVPs should be processed. The lowest item numbers are processed first.

SIP-Number-Auth-Items AVP This AVP is used in a request to indicate the number of authentication vectors being requested by the S-CSCF. The Diameter server (such as the HSS) then provides the authentication vectors using this AVP in the answer command.

Subscription-Info AVP   This grouped AVP is used to send subscription information for the specified device.

<Subscription-Info>::=<AVP Header: 642 10415>

{Call-ID-SIP-Header}

{From-SIP-Header}

{To-SIP-Header}

{Record-Route}

{Contact}

*[AVP]

Supported-Applications AVP   This is a grouped AVP that identifies the supported applications of a Diameter network element.

<Supported-Applications>::=<AVP Header: 631 10415>

*[Auth-Application-ID]

*[Acct-Application-ID]

*[Vendor-Specific-Application-ID]

*[AVP]

Supported-Features AVP   This is a grouped AVP used to indicate the features supported by a destination host. This AVP is commonly used when a session between two network nodes is being established to ensure the destination is able to support the required features of the originator, and to ensure the originator does not request services of the destination that are not supported.

<Supported-Features>::=<AVP Header: 628 10415>

{Vendor-ID}

{Feature-List-ID}

{Feature-List}

*[AVP]

To-SIP-Header AVP   This AVP contains the content of the To SIP header and is used in the Subscription-Info AVP.

UAR-Flags AVP   This bit-masked AVP is used to identify a session as an IMS emergency registration request (Table 8.9).

User-Authorization-Type AVP   This is an enumerated AVP used to identify the type of authorization being used for a user.

0 = Registration. This is determined by the I-CSCF by looking at the Expires SIP parameter in the SIP REGISTER. This is the default value.

Bit	Description
0	When this bit is set, it identifies the request as an IMS emergency registration request.

TABLE 8.9 UAR-Flags Value for Cx

1 = Deregistration. This is determined by the I-CSCF by looking at the Expires SIP parameter in the SIP REGISTER.

2 = Registration and capabilities. This gets used when the I-CSCF requests capability information for the S-CSCF from the HSS.

User-Data-Already-Available AVP This is an enumerated AVP used by the HSS to verify if the S-CSCF has all the information from the user profile that is needed to provide service to the subscriber.

0 = User data not available

1 = User data already available

User-Data AVP The user data is the subscription information for any one subscriber. The exact formatting for the data can be found in Annex B of 3GPP TS 29.228.

Visited-Network-Identifier AVP This AVP is used by the HSS to uniquely identify the visited network. The identity is derived by the I-CSCF when it receives the P-Visited-Network-ID. The coding used in the octet string is unique to the home network.

Wildcarded-IMPU AVP This AVP is used to send a wildcarded public user identity from the HSS. It is used on the Sh interface.

Wildcarded-Public-Identity AVP This AVP is used to send a wildcarded public identity that is stored in the HSS.

Acronyms

2G	Second generation wireless
3G	third generation wireless
3GPP	Third Generation Partnership Project
4G	fourth generation wireless
5G	fifth generation wireless
ADC	Application Detection and Control
AF	application function
AMBR	aggregate maximum bit rate
ANDSF	access network discovery and selection function
APN	access point name
AuC	authentication center
AVP	attribute value pair
BBERF	Bearer Binding and Event Reporting Function
BCM	bearer control mode
CAMEL	Customized Applications for Mobile Enhanced Logic
CCPCH	Common Control Physical CHannel (for UMTS)
CDF	charging data function
CDMA	Code Division Multiple Access
CDR	call detail record
CI	cell identity
CPICH	Common PIlot CHannel (for UMTS)
CSCF	call session control function
CSG	closed subscriber group
CTF	charging trigger function
DEA	Diameter edge agent
DPI	deep packet inspection
DSC	Diameter signaling controller
DSCP	Differentiated Services Code Point
DTLS	Datagram Transport Layer Security
ECGI	EUTRAN Cell Global ID
EIR	equipment identity register

eNode B	Evolved Node B
EPC	evolved packet core
EPS	evolved packet system
ETSI	European Telecommunications Standards Institute
FBA	fixed broadband access
FDD	Frequency Division Duplex
FQDN	fully qualified domain name
GETS	Government Emergency Telephone Service
GGSN	gateway GPRS support node
GIBA	GPRS IMS Bundled Authentication
GMLC	gateway Mobile Location Center
GSM	Global System for Mobile
GTP	GPRS Tunneling Protocol
HeNB	Home enhanced Node B
HLR	home location register
HSS	Home Subscriber Server
IANA	Internet Assigned Numbers Authority
I-CSCF	Interrogating-Call Session Control Function
iFC	initial Filter Criteria
IMEI	International Mobile Equipment Identity
IMPU	IP Multimedia Public Identity
IMS	IP Multimedia Subsystem
IMSI	International Mobile Subscriber Identity
IoT	Internet of Things
IP-CAN	IP connectivity access network
ITU	International Telecommunications Union
LAI	location area identification
LIPA	local IP access
LTE	Long Term Evolution
M2M	machine-to-machine
MDT	minimalized drive test
MIP	mobile IP
MLC	Mobile Location Center
MME	Mobility Management Entity
MSC	Mobile Switching Center
MSISDN	Mobile Subscriber ISDN

NAT	network address translation
NGN	next generation network
NTP	Network Time Protocol
OCF	online charging, function
OCS	online charging system
OFCS	offline charging system
PCC	policy and charging control
PCEF	policy control enforcement function
PCRF	policy and charging rules function
P-CSCF	Proxy-Call Session Control Function
PDN	packet data network
PDP	packet data protocol
PGW	packet data network gateway
PLMN	public land mobile network
PPP	Point-to-Point Protocol
ProSe	proximity-based services
PSI	Public Service Identity
QCI	QoS Class Identifier
QoS	quality of service
RAB	Radio Access Bearer
RADIUS	Remote Authentication Dial-In User Service
RAN	radio access network
RAND	RANDom number
RAT	radio access type
RAU	routing area update
RNC	radio network controller
RSCP	Received Signal Code Power
RSRP	Reference Signal Received Power
RSRQ	Reference Signal Received Quality
RTCP	RTP Control Protocol
RTP	Real Time Protocol
RTWP	Received Total Wideband Power
SCCP	Signaling Connection Control Part
S-CSCF	Serving-Call Session Control Function
SCTP	Stream Control Transmission Protocol
SDP	Session Description Protocol

SEG	security gateway
SGSN	serving GPRS support node
SGW	serving gateway
SiFC	Shared initial Filter Criteria
SIGTRAN	SIGnaling Transport
SIP	Session Initiation Protocol
SIPTO	Selected IP traffic offload
SLF	subscriber location function
SPI	security parameter index
SPR	subscriber profile repository
SRVCC	Single Radio Voice Call Continuity
SS7	Signaling System #7
STN	session transfer number
STP	signal transfer point
TAC	type allocation code
TAI	Tracking Area Identity
TAU	tracking area update
TCP	Transmission Control Protocol
TDD	time division duplex
TDF	traffic detection function
TDM	time division multiplex
TFT	traffic flow template
TLS	Transport Layer Security
UDC	User Data Convergence
UDP	User Datagram Protocol
UDR	user data repository
UE	user equipment
UMTS	Universal Mobile Telecommunications System
VLR	visitor location register
VoIP	Voice over IP
VoLTE	Voice over Long Term Evolution

Bibliography

"GPRS: General Packet Radio Service;" Bates, Regis "Bud" J.; McGraw-Hill, NY 2002.

3GPP TS 20.003: "Numbering, Addressing and Identification."

3GPP TS 21.133: "3G Security; Security Threats and Requirements."

3GPP TS 22.004: "General on supplementary services."

3GPP TS 22.173: "IP Multimedia Core Network Subsystem (IMS) Multimedia Telephony Service and supplementary services; Stage 1."

3GPP TS 22.803: "Feasibility study for Proximity Services (ProSe)."

3GPP TS 23.003: "Numbering, addressing and identification."

3GPP TS 23.203: "Policy and charging control architecture."

3GPP TS 23.216: "Single Radio Voice Call Continuity (SRVCC); Stage 2."

3GPP TS 23.246: "Multimedia Broadcast/Multicast Service (MBMS); Architecture and functional description."

3GPP TS 23.271: "Functional stage 2 description of Location Services (LCS)."

3GPP TS 23.292: "IP Multimedia Subsystem (IMS) Centralized Services; Stage 2."

3GPP TS 23.401: "General Packet Radio Service (GPRS) enhancements for Evolved Universal Terrestrial Radio Access Network (E-UTRAN) access."

3GPP TS 23.402: "Architecture enhancements for non-3GPP accesses."

3GPP TS 23.468: "LTE; Group Communication System Enablers for LTE (GCSE_LTE); Stage 2."

3GPP TS 24.008: "Mobile radio interface Layer 3 specification; Core network protocols; Stage 3."

3GPP TS 25.331: "Radio Resource Control (RRC); Protocol specification."

3GPP TS 25.992: "Multimedia Broadcast/Multicast Service (MBMS); UTRAN/GERAN requirements."

3GPP TS 29.002: "Mobile Application Part (MAP) specification."

3GPP TS 29.060: "General Packet Radio Service (GPRS); GPRS Tunnelling Protocol (GTP) across the Gn and Gp interface."

3GPP TS 29.061: "Universal Mobile Telecommunications System (UMTS); LTE; Interworking between the Public Land Mobile Network (PLMN) supporting packet based services and Packet Data Networks (PDN)."

3GPP TS 29.172: "Location Services (LCS); Evolved Packet Core (EPC) LCS Protocol (ELP) between the Gateway Mobile Location Centre (GMLC) and the Mobile Management Entity (MME); SLg interface."

3GPP TS 29.173: "Location Services (LCS); Diameter-based SLh interface for Control Plane LCS."

3GPP TS 29.212: "Policy and Charging Control (PCC); Reference points."

3GPP TS 29.213: "Policy and Charging Control signalling flows and Quality of Service (QoS) parameter mapping."

3GPP TS 29.214: "Policy and Charging Control over Rx reference point."

3GPP TS 29.219: "Policy and Charging Control: Spending Limit Reporting over Sy reference point."

3GPP TS 29.228: "IP Multimedia (IM) Subsystem Cx and Dx interfaces; Signalling flows and message contents."

3GPP TS 29.229: "Cx and Dx interfaces based on the Diameter protocol; Protocol details."

3GPP TS 29.230: "Diameter applications; 3GPP specific codes and identifiers."

3GPP TS 29.272: "Evolved Packet System (EPS); Mobility Management Entity (MME) and Serving GPRS Support Node (SGSN) related interfaces based on Diameter protocol."

3GPP TS 29.274: "3GPP Evolved Packet System (EPS); Evolved General Packet Radio Service (GPRS) Tunnelling Protocol for Control plane (GTPv2-C); Stage 3."

3GPP TS 29.305: "InterWorking Function (IWF) between MAP based and Diameter based interfaces."

3GPP TS 29.328: "IP Multimedia (IM) Subsystem Sh interface; Signalling flows and message contents."

3GPP TS 29.329: "Sh Interface based on the Diameter protocol; Protocol details."

3GPP TS 29.364: "IP Multimedia Subsystem (IMS) Application Server (AS) service data descriptions for AS interoperability."

3GPP TS 32.299: "Telecommunication management; Charging management; Diameter charging applications."

3GPP TS 32.422: "Telecommunication management; Subscriber and equipment trace; Trace control and configuration management."

3GPP TS 33.102: "3G Security; Security architecture."

3GPP TS 33.210: "3G Security; Network Domain Security; IP network layer security."

3GPP TS 37.320: "Universal Terrestrial Radio Access (UTRA) and Evolved Universal Terrestrial Radio Access (E-UTRA); Radio measurement collection for Minimization of Drive Tests (MDT); Overall description; Stage 2."

ETSI ES 283 035 V3.1.1: "Network Technologies (NTECH); Network Attachment; e2 interface based on the DIAMETER protocol."

Olson, M., 2009. "SAE and the Evolved Packet Core." Academic Press, ISBN: 978-0-12-374826-3.

RFC 2924: "Accounting Attributes and Record Formats."

RFC 2975: "Introduction to Accounting Management."

RFC 3162: "RADIUS and IPv6."

RFC 3261: "SIP: Session Initiation Protocol."

RFC 3327: "Session Initiation Protocol (SIP) Extension Header Field for Registering Non-Adjacent Contacts."

RFC 3539: "Authentication, Authorization and Accounting (AAA) Transport Profile."

RFC 3554: "On the Use of Stream Control Transmission Protocol (SCTP) with IPsec."

RFC 3588: "Diameter Base Protocol."

RFC 4005: "Diameter Network Access Server Application."

RFC 4006: "Diameter Credit-Control Application."

RFC 4282: "The Network Access Identifier."

RFC 4740: "Diameter Session Initiation Protocol (SIP) Application."

RFC 5234: "Augmented BNF for Syntax Specifications: ABNF."

RFC 5246: "The Transport Layer Security (TLS) Protocol Version 1.2."

RFC 5729: "Clarifications on the Routing of Diameter Requests Based on the Username and the Realm."

RFC 6083: "Datagram Transport Layer Security (DTLS) for Stream Control Transmission Protocol (SCTP)."

RFC 6733: "Diameter Base Protocol."

RFC 7155: "Diameter Network Access Server Application."

RFC 7423: "Diameter Applications Design Guidelines."

Index

Note: Page numbers followed by an *f* and a *t* indicate figure and table, respectively.

www.ingramcontent.com/pod-product-compliance
Lightning Source LLC
Chambersburg PA
CBHW061923190326
41458CB00009B/2639